T0360595

Systems Engineering Using the DEJI Systems Model®

While we need to work more with a systems approach, there are few books that provide systems engineering theory and applications. This book presents a comprehensive collection of systems engineering models. Each of the models is fully covered with guidelines on how and why to use them, along with case studies.

Systems Engineering Using the DEJI Systems Model®: Evaluation, Justification, and Integration with Case Studies and Applications provides systems integration as a unifying platform for systems of systems and presents a structured model for systems applications and explicit treatment of human-in-the-loop systems. It discusses systems design in detail and covers the justification methodologies along with examples. Systems evaluation tools and techniques are also included with a discussion on how engineering education is playing a major role for systems advancement.

Practicing professionals, as well as educational institutions, governments, businesses, and industries, will find this book of interest.

Systems Innovation Book Series

Series Editor: Adedeji Badiru

Systems Innovation refers to all aspects of developing and deploying new technology, methodology, techniques, and best practices in advancing industrial production and economic development. This entails such topics as product design and development, entrepreneurship, global trade, environmental consciousness, operations and logistics, introduction and management of technology, collaborative system design, and product commercialization. Industrial innovation suggests breaking away from the traditional approaches to industrial production. It encourages the marriage of systems science, management principles, and technology implementation. Particular focus will be the impact of modern technology on industrial development and industrialization approaches, particularly for developing economics. The series will also cover how emerging technologies and entrepreneurship are essential for economic development and society advancement.

Project Management
Systems, Principles, and Applications, Second Edition
Adedeji B. Badiru

Global Manufacturing Technology Transfer
Africa-USA Strategies, Adaptations, and Management
Adedeji B. Badiru

Data Analytics
Handbook of Formulas and Techniques
Adedeji B. Badiru

Conveyors
Application, Selection, and Integration
Patrick M McGuire

Innovation Fundamentals
Quantitative and Qualitative Techniques
Adedeji B. Badiru and Gary Lamont

Global Supply Chain
Using Systems Engineering Strategies to Respond to Disruptions
Adedeji B. Badiru

Systems Engineering Using the DEJI Systems Model®
Evaluation, Justification, and Integration with Case Studies and Applications
Adedeji B. Badiru

Systems Engineering Using the DEJI Systems Model®

Evaluation, Justification, and Integration
with Case Studies and Applications

Adedeji B. Badiru

CRC Press
Taylor & Francis Group
Boca Raton London New York

CRC Press is an imprint of the
Taylor & Francis Group, an **informa** business

MATLAB® is a trademark of The MathWorks, Inc. and is used with permission. The MathWorks does not warrant the accuracy of the text or exercises in this book. This book's use or discussion of MATLAB® software or related products does not constitute endorsement or sponsorship by The MathWorks of a particular pedagogical approach or particular use of the MATLAB® software.

First edition published 2023
by CRC Press
6000 Broken Sound Parkway NW, Suite 300, Boca Raton, FL 33487-2742

and by CRC Press
4 Park Square, Milton Park, Abingdon, Oxon, OX14 4RN

CRC Press is an imprint of Taylor & Francis Group, LLC

ISBN: 978-1-032-00802-8 (hbk)
ISBN: 978-1-032-00803-5 (pbk)
ISBN: 978-1-003-17579-7 (ebk)

DOI: 10.1201/9781003175797

Typeset in Times
by SPi Technologies India Pvt Ltd (Straive)

Dedication

Dedicated to the memory of Professor Oyewusi Ibidapo-Obe, who dedicated his career to systems implementations, even in the face of the stochasticity of a chaotic world.

Contents

Preface

As an educator, researcher, scholar, and author, I see everything from the lens of a system. It is a systems World and we must think and act accordingly. It is in that context that the idea for this book evolved.

Enthused by analytical and computer tools at their disposal, technical professionals tend to jump to the coordination and implementation stages of a system or project. That is, jumping to the design functionality stage while dispensing with intermediate steps, where non-technical and "soft" issues might exist. However, those intermediate stages are often more critical for systems success than the pure analytical foundation. Items such as needs analysis, gap analysis, user involvement, communication, cooperation, resource requirement analysis, budget flow, leadership support, workforce acceptance, and system desirability are essential before getting to the implementation stage (end point) of the project. This is the essential narrative that highlights the efficacy of the approach of DEJI Systems Model®, which takes a system sequentially through the stages of design, evaluation, justification, and, finally, integration. These step-by-step stages allow important considerations, technical or otherwise, to be addressed in the project. It is of utmost importance to understand how the expected output of the project will integrate with and align with the existing organizational framework. DEJI systems model makes it imperative to do an a priori evaluation of the potential impact that the project output might have on the prevailing operating environment.

Adedeji B. Badiru

Acknowledgments

Several people contributed in various ways to the compilation of the manuscript for this book, in terms of suggestions, recommendations, editorial support, and case studies. Of particular importance are the case studies provided by Mike Kaminski, of Accenture Federal Services, on the practical applications of the DEJI Systems Model® in process-improvement initiatives. I also express my extreme gratitude to John and Abi Egan, who provided a quiet and secluded writing space in their home during my writing retreat in Dallas, Texas, in September 2021. Without that focused writing space, this manuscript could not have been completed as rapidly as it was.

Author

Adedeji B. Badiru is professor of Systems Engineering and dean of the Graduate School of Engineering and Management at the Air Force Institute of Technology (AFIT). He has oversight of planning, directing, and controlling operations related to granting doctoral and master's degrees, professional continuing cyber education, and research and development programs for the US Air Force. He was previously professor and head of Systems Engineering and Management at AFIT; professor and department head of Industrial Engineering at the University of Tennessee, Knoxville; and professor of Industrial Engineering and dean of University College at the University of Oklahoma, Norman. He is a registered professional engineer (PE), a certified project management professional (PMP), a fellow of the Institute of Industrial & Systems Engineers, and a fellow of the Nigerian Academy of Engineering. He is also a program evaluator (PEV) for ABET. He holds a leadership certificate from the University Tennessee Leadership Institute. He has BS in Industrial Engineering, MS in Mathematics, and MS in Industrial Engineering from Tennessee Technological University and PhD in Industrial Engineering from the University of Central Florida. His areas of interest include mathematical modeling, project modeling and analysis, economic analysis, systems engineering modeling, computer simulation, and productivity analysis. He is a prolific author, with over 35 books, over 30 book chapters, over 130 Journal and magazine articles, and over 200 conference presentations. He is a member of several professional associations and scholastic honor societies. Professor Badiru, a world-renowned educator, has won several awards for his teaching, research, administrative, and professional accomplishments. Some of his selected awards include the 2009 Dayton Affiliate Society Council Award for Outstanding Scientists and Engineers in the Education category with a commendation from the 128th Senate of Ohio, 2010 ASEE John Imhoff Award for his global contributions to Industrial Engineering Education, the 2011 Federal Employee of the Year Award in the Managerial Category from the International Public Management Association, Wright Patterson Air Force Base, the 2012 Distinguished Engineering Alum Award from the University of Central Florida, the 2012 Medallion Award from the Institute of Industrial Engineers for his global contributions in the advancement of the profession, 2016 Outstanding Global Engineering Education Award from the Industrial Engineering and Operations Management (IEOM), 2015 Air Force-level Winner of the National Public Service Award from The American Society for Public Administration and the National Academy of Public Administration, 2013 Father D. J. Slattery Excellence Award, Saint Finbarr's College Alumni Association, North America Chapter, 2013 Award Team Leader, Air Force Organizational Excellence Award for Air University C3 (Cost Conscious Culture), 2013 Finalist for Jefferson Science Fellows Program, National Academy of Sciences, 2012 Book-of-the-Month Recognition for Statistical Techniques for Project Control from the Industrial Engineering

Magazine, and the 2010 Industrial Engineering Joint Publishers Book-of-the-Year Award for The Handbook of Military Industrial Engineering.

Professor Badiru is also the book series editor for CRC Press/Taylor & Francis Group book series on Systems Innovation. He has served as a consultant to several organizations around the world, including Russia, Mexico, Taiwan, Nigeria, and Ghana. He has conducted customized training workshops for numerous organizations, including Sony, AT&T, Seagate Technology, U.S. Air Force, Oklahoma Gas & Electric, Oklahoma Asphalt Pavement Association, Hitachi, Nigeria National Petroleum Corporation, and ExxonMobil. He has served as a technical project reviewer, curriculum reviewer, and proposal reviewer for several organizations, including The Third-World Network of Scientific Organizations, Italy, Social Sciences and Humanities Research Council of Canada, National Science Foundation, National Research Council, and the American Council on Education. He is on the editorial and review boards of several technical journals and book publishers. Professor Badiru has also served as an industrial development consultant to the United Nations Development Program. In 2011, Professor Badiru led a research team to develop analytical models for Systems Engineering Research Efficiency (SEER) for the Air Force acquisitions integration office at the Pentagon. He has led a multi-year multi-million dollar research collaboration between the Air Force Institute of Technology and KBR Aerospace Group. Professor Badiru has diverse areas of avocation. His professional accomplishments are coupled with his passion for writing about everyday events and interpersonal issues, especially those dealing with social responsibility. Outside of the academic realm, he writes motivational poems, editorials, and newspaper commentaries, as well as engages in paintings and crafts. Professor Badiru is the 2020 recipient of the Lifetime Achievement Award from Taylor & Francis publishing group. He was also part of the AFIT team that led the institution's receipt of the 2019/2020 US Air Force Organizational Excellence Award. Dr. Badiru is also the recipient of the 2022 BEYA career achievement award in the Government category.

1 Systems View of the World

INTRODUCTION

It is not enough to proclaim a systems view of the world. We must actualize and practicalize the essence of a systems-centric world. At the end of the day, integration, in the context of the DEJI Systems Model presented in this book, is the critical key for actualizing a systems view. It is a proven fact that methodologies imported from one cultural system into another cultural system don't work well until they are adapted and integrated into the prevailing local practices of the target system. This can be observed around our present world, with vastly differing platforms of political systems, cultural systems, financial systems, educational systems, media systems, international trade systems, and other world-affective systems. In as much as the world is expected to work together, we must be cognizant of our cultural and operational differences. We must not be dismissive of where disconnections exist. Rather, we should confront, address, mitigate, or rectify the dissimilarities as we go through the structured tenets of the DEJI Systems Model of design, evaluation, justification, and, finally, integration. If this process is embraced and imbibed at the outset, from a systems perspective, we will have a better opportunity to make things (and people) work productively together in the long run. Figure 1.1 presents a pictorial representation of why integration (by any other name) is essential for achieving the ideals of a cohesive system. It is not expected that systems will align perfectly in all respects at all times. However, to whatever extent possible, we need to maximize the areas of overlap. This can be done through a combination of qualitative and quantitative techniques.

ALIGNMENT WITH THE TENETS OF IEOM

The premise of this book is to present a motivating spread of soft and hard tools and techniques for systems integration. It is on this duality of soft and hard approaches that operational alignment is of particular interest to the Industrial Engineering and Operations Management (IEOM) Society (ieomsociety.org). This commitment permeates all the principles of IEOM, as the organization works inclusively to align and integrate professional, operational, and educational interests around the world. The vision, mission, and values of IEOM are echoed below to put this book in an operational context.

DOI: 10.1201/9781003175797-1

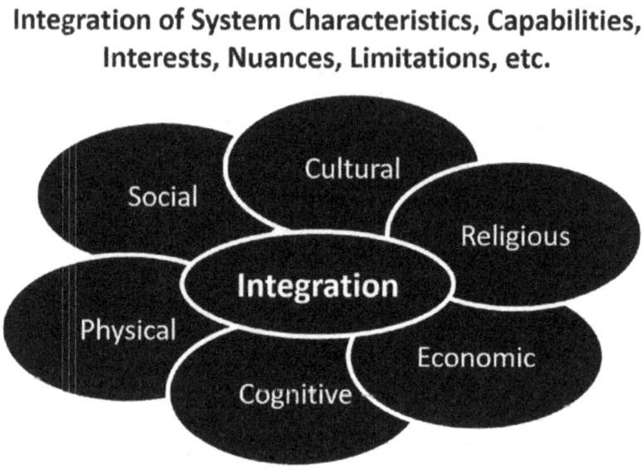

Integration of System Characteristics, Capabilities, Interests, Nuances, Limitations, etc.

FIGURE 1.1   Integration of system characteristics across the spectrum of world affairs.

IEOM Vision

The IEOM Society International strives to be the premier global organization dedicated to the advancement of industrial engineering and operations management discipline for the betterment of humanity.

IEOM Mission

IEOM Society's core purpose is to globally foster critical thinking and its effective utilization in the field of industrial engineering (IE) and operations management (OM) by providing means to communicate and network among diversified people, especially in emerging countries, motivated by similar interests.

IEOM Values

- Globalization, diversity, and inclusion
- Innovative and entrepreneurial thinking
- Student empowerment and mentoring
- Professional development and lifelong learning
- Building partnerships among industry, academia, and government
- Sustainability and community development

A systems view of the world is required for people to live and work together for a common good around the world. The profession of industrial engineering can help actualize a systems view of the world. It is in this context and other operational strategies that the former Institute of Industrial Engineers (IIE) changed its name in 2016 to the Institute of Industrial and Systems Engineering (IISE). For decades, the institute was known as the American Institute of Industrial Engineers

(AIIE). It was expanded in 1981 by dropping the "American" component of the name in recognition of the more international scope of the profession. The evolution of the profession over the years conveys the growing realization of a systems view of the world, beyond the traditional in-ward looking perspectives of organizations. Founded in 1948, IISE is the only international, nonprofit, professional society dedicated to advancing the technical and managerial excellence of industrial engineers. The profession of industrial engineering is recognized and applauded for advancing the practice of systems view globally. This is noted enough to imbibe this author's cliché that "things operate better when industrial engineers are involved."

In the final analysis, everything works as a system. It is a systems world and everything around us is driven by a systems underpinning of how things work, both individually and together. Systems thinking facilitates the incorporation of all the elements and nuances in the operating environment. Nothing typifies this fact more than the profession of industrial and systems engineering (ISE). The foundational basis of industrial engineering is the appreciation of how "systems" permeates everything we do in business, industry, government, academia, the military, and other enterprises.

In response to the emergence of the COVID-19 pandemic in 2020, the world has learned to demonstrate, accept, embrace, and practice more flexibility in work, leisure, personal, and family activities. This calls for a more integrative systems view of everything we do. Under a systems perspective, whatever personal choices we make end up directly affecting other citizens. This realization of herd-to-herd systems interactions may help us nudge along those who are resistant, hesitant, or obstinate about community health initiatives.

It is the systems orientation that gives industrial engineering the flexibility, versatility, and professional mobility that it has. This is encapsulated in the very definition of the profession, as echoed in the section that follows.

INDUSTRIAL ENGINEERING AND SYSTEMS APPROACH

A common question is "who is an industrial engineer?" Because the profession of industrial engineering is very diverse, flexible, and widely encompassing, it is often challenging to define the wide scope of industrial engineering in a few sentences.

WHO IS AN INDUSTRIAL ENGINEER?

An Industrial Engineer is someone who is concerned with the design, installation, and improvement of integrated systems of people, materials, information, equipment, and energy by drawing upon specialized knowledge and skills in the mathematical, physical, and social sciences, together with the principles and methods of engineering analysis and design to specify, predict, and evaluate the results to be obtained from such systems.

INDUSTRIAL ENGINEERING AND INNOVATION

Systems implementation drives innovation. Thus, by inference, industrial engineering principles are foundational for driving, anchoring, and sustaining innovation, from a systems platform. Whatever your industry or business, you are operating in a system. The comprehensive definition of industrial engineering says it aptly. In the perspective of industrial engineering, everything is a system. The more things are viewed as a system, the better we can have a consistent and comprehensive handle on its operations. ISEs are perhaps the most preferred engineering professionals because of their ability to manage complex organizations (Oke, 2014). They are trained to design, develop, and install optimal methods for coordinating people, materials, equipment, energy, and information. The integration of these resources is needed in order to create products and services in a business world that is becoming increasingly complex and globalized. Industrial and systems engineers oversee management goals and operational performance. Their aims are the effective management of people, coordinating techniques in business organization, and adopting technological innovations toward achieving increased performance. They also stimulate awareness of the legal, environmental, and socioeconomic factors that have a significant impact on engineering systems. Industrial and systems engineers can apply creative values in solving complex and unstructured problems in order to synthesize and design potential solutions and organize, coordinate, lead, facilitate, and participate in teamwork. They possess good mathematical skills, a strong desire for organizational performance, and a sustained drive for organizational improvement.

In deriving efficient solutions to manufacturing, organizational, and associated problems, ISEs analyze products and their requirements. They utilize mathematical techniques such as operations research (OR) to meet those requirements and to plan production and information systems. They implement activities to achieve product quality, reliability, and safety by developing effective management control systems to meet financial and production planning needs. Systems design and development for the continual distribution of the product or service is also carried out by ISEs to enhance an organization's ability to satisfy their customers. Industrial and systems engineers focus on optimal integration of raw materials available, transportation options, and costs in deciding plant location. They coordinate various activities and devices on the assembly lines through simulations and other applications.

The organization's wage and salary administration systems and job evaluation programs can also be developed by them, leading to their eventual absorption into management positions. They share similar goals with health and safety engineers in promoting product safety and health in the whole production process through the application of knowledge of industrial processes and such areas as mechanical, chemical, and psychological principles. They are well grounded in the application of health and safety regulations while anticipating, recognizing, and evaluating hazardous conditions and developing hazard-control techniques.

Industrial and systems engineers can assist in developing efficient and profitable business practices by improving customer services and the quality of products. This would improve the competitiveness and resource utilization in organizations. From another perspective, ISEs are engaged in setting traditional labor or time standards and in redesigning organizational structure in order to eliminate or reduce some forms of frustration or wastes in manufacturing. This is essential for the long-term survivability and the health of the business.

Another aspect of the business that the ISEs could be useful in is making work safer, easier, more rewarding, and faster through better designs that reduce production cost and allow the introduction of new technologies. This improves the lifestyle of the populace by making it possible for them to afford and use technologically advanced goods and services. In addition, they offer ways of improving the working environment, thereby improving efficiencies and increasing cycle time and throughput, and helping manufacturing organizations to obtain their products more quickly. Also, ISEs have provided methods by which businesses can analyze their processes and try to make improvements to them. They focus on optimization—doing more with less—and help to reduce waste in the society. The ISEs give assistance in guiding the society and businesses to care more for their workforce while improving the bottom line.

Since this handbook deals with two associated fields—industrial and systems engineering—there is a strong need to define these two professions in order to have a clear perspective about them and to appreciate their interrelationships. Throughout this chapter, these two fields are used together and the discussions that follow are applicable to either. Perhaps the first classic and widely accepted definition of industrial engineering (IE) was offered by the then AIIE in 1948. Others have extended the definition. "Industrial Engineering is uniquely concerned with the analysis, design, installation, control, evaluation, and improvement of sociotechnical systems in a manner that protects the integrity and health of human, social, and natural ecologies. A sociotechnical system can be viewed as any organization in which people, materials, information, equipment, procedures, and energy interact in an integrated fashion throughout the life cycles of its associated products, services, or programs. Through a global system's perspective of such organizations, industrial engineering draws upon specialized knowledge and skills in the mathematical, physical, and social sciences, together with the principles and methods of engineering analysis and design, to specify the product and evaluate the results obtained from such systems, thereby assuring such objectives as performance, reliability, maintainability, schedule adherence, and cost control.

There are five general areas of industrial and systems engineering. Each of these areas specifically makes out some positive contributions to the growth of industrial and systems engineering. The first area shown in the diagram is twofold and comprises sociology and economics. The combination of the knowledge from these two areas helps in the area of supply chain. The second area is mathematics, which is a powerful tool of ISEs. Operations research is an important part of this area. The third area is psychology, which is a strong pillar of ergonomics. Accounting and economics both constitute the fourth area. These are useful subjects in the area of engineering economics. The fifth area is computer. Computers are helpful in CAD/CAM, which is an important area of industrial and systems engineering.

According to the International Council on Systems Engineering (INCOSE), systems engineering is an interdisciplinary approach and means to enable the realization of successful systems. Such systems can be diverse, encompassing people and organizations, software and data, equipment and hardware, facilities and materials, and services and techniques. The system's components are interrelated and employ organized interaction toward a common purpose. From the viewpoint of INCOSE, systems engineering focuses on defining customer needs and required functionality early in the development cycle, documenting requirements, and then proceeding with design synthesis and systems validation while considering the complete problem. The philosophy of systems engineering teaches that attention should be focused on what the entities do before determining what the entities are. A good example to illustrate this point may be drawn from the transportation system. In solving a problem in this area, instead of beginning the problem-solving process by thinking of a bridge and how it will be designed, the systems engineer is trained to conceptualize the need to cross a body of water with certain cargo in a specific way.

The systems engineer then looks at bridge design from the point of view of the type of bridge to be built. For example, is it going to have a suspension or superstructure design? From this stage he would work down to the design detail level where systems engineer gets involved, considering foundation soil mechanics and the placement of structures. The contemporary business is characterized by several challenges. This requires the ISEs to have skills, knowledge, and technical know-how in the collection, analysis, and interpretation of data relevant to problems that arise in the workplace. This places the organization well above the competition.

The radical growth in global competition, constantly and rapidly evolving corporate needs, and the dynamic changes in technology are some of the important forces shaping the world of business. Thus, stakeholders in the economy are expected to operate within a complex but ever-changing business environment. Against this backdrop, the dire need for professionals who are reliable, current, and relevant becomes obvious. Industrial and systems engineers are certainly needed in the economy for bringing about radical change, value creation, and significant improvement in productive activities.

The ISE must be focused and have the ability to think broadly in order to make a unique contribution to the society. To complement this effort, the organization itself must be able to develop effective marketing strategies (aided by a powerful tool, the Internet) as a competitive advantage so that the organization could position itself as the best in the industry.

The challenges facing the ISE may be divided into two categories: those faced by ISEs in developing and underdeveloped countries and those faced by engineers in developed countries. In developed countries, there is a high level of technological sophistication that promotes and enhances the professional skills of the ISE.

Unfortunately, the reverse is the case in some developing and underdeveloped countries. Engineers in underdeveloped countries, for instance, rarely practice technological development, possibly owing to the high level of poverty in such environments. Another reason that could be advanced for this is the shortage of skilled manpower in the engineering profession that could champion technological breakthrough similar to the channels operated by the world economic powers. In addition, the technological development of nations could be enhanced by the formulation of active research teams. Such teams should be focused with the aims of solving practical industrial problems. Certain governments in advanced countries encourage engineers (including ISEs) to actively participate in international projects funded by government or international agencies. For the developing and underdeveloped countries, this benefit may not be gained by the ISE until the government is challenged to do so in order to focus on the technological development of the country.

Challenges before a community may be viewed from the perspective of the problem faced by the inhabitants of that community. As such, they could be local or global. Local challenges refer to the need that must be satisfied by the engineers in that community. These needs may not be relevant to other communities; for example, the ISE may be in a position to advise the local government chairman of a community on the disbursement of funds on roads within the powers of the local government. Decision-science models could be used to prioritize certain criteria, such as the number of users, the economic indexes of the various towns and villages, the level of business activities, the number of active industries, the length of the road, and the topography or the shape of the road.

Soon after graduation, an ISE is expected to tackle a myriad of social, political, and economic problems. This presents a great challenge to the professionals who live in a society where these problems exist. Consider the social problems of electricity generation, water provision, flood control, etc. The ISE in a society where these problems exist is expected to work together with other engineers in order to solve these problems. They are expected to design, improve on existing designs, and install integrated systems of men, materials, and equipment so as to optimize the use of resources. For electricity distribution, the ISE should be able to develop scientific tools for the distribution of power generation as well as for the proper scheduling of the maintenance tasks to which the facilities must be subjected.

The distribution network should minimize the cost. Loss prevention should be a key factor to consider. As such, the quality of the materials purchased for maintenance should be controlled, and a minimum acceptable standard should be established. In solving water problems, for instance, the primary distribution route should be a major concern. The ISE may need to develop reliability models that could be applied to predict the life of components used in the system. The scope of activities of the ISE should be wide enough for them to work with other scientists in the health sector on modeling and control of diseases caused by water-distribution problems. The ISE should be able to solve problems under uncertain conditions and limited budgets.

The ISE can work in a wide range of industries, such as the manufacturing, logistics, service, and defense industries. In manufacturing, the ISE must ensure that the equipment, manpower, and other resources in the process are integrated in such a manner that efficient operation is maintained and continuous improvement is ensured. The ISE functions in the logistics industry through the management of supply-chain systems (e.g., manufacturing facilities, transportation carriers, distribution hubs, retailers) to fulfill customer orders in the most cost-effective way. In the service industry, the ISE provides consultancies in areas related to organizational effectiveness, service quality, information systems, project management, banking, service strategy, etc. In the defense industry, the ISE provides tools to support the management of military assets and military operations in an effective and efficient manner. The ISE works with a variety of job titles. The typical job titles of an ISE graduate include industrial engineer, manufacturing engineer, logistics engineer, supply-chain engineer, quality engineer, systems engineer, operations analyst, management engineer, and management consultant.

Experiences in the United States and other countries show that a large proportion of ISE graduates work in consultancy firms or as independent consultants, helping companies to engineer processes and systems to improve productivity, effect efficient operation of complex systems, and manage and optimize these processes and systems.

After completing their university education, ISEs acquire skills from practical exposure in an industry. Depending on the organization that an industrial or systems engineer works for, the experience may differ in depth or coverage. The trend of professional development in industrial and systems engineering is rapidly changing in recent times. This is enhanced by the ever-increasing development in the information, communication and technology (ICT) sector of the economy.

Industrial and systems engineering is methodology-based and is one of the fastest-growing areas of engineering. It provides a framework that can be focused on any area of interest, and incorporates inputs from a variety of disciplines, while maintaining the engineer's familiarity and grasp of physical processes. The honor of discovering industrial engineering belongs to a large number of individuals. The eminent scholars in industrial engineering are Henry Gantt (the inventor of the Gantt chart) and Lillian Gilbreth (a coinventor of time and motion studies). Some other scientists have also contributed immensely to its growth over the years. The original application of industrial engineering at the turn of the century was in manufacturing a technology-based orientation, which gradually changed with the development of OR, cybernetics, modern control theory, and computing power.

Computers and information systems have changed the way industrial engineers do business. The unique competencies of an ISE can be enhanced by the powers of the computer. Today, the fields of application have widened dramatically, ranging from the traditional areas of production engineering, facilities planning, and material handling to the design and optimization of more broadly defined systems. An ISE is a versatile professional who uses scientific tools in problem solving through a holistic and integrated approach. The main objective of an ISE is to optimize performance through the design, improvement, and installation of integrated human, machine, and equipment systems. The uniqueness of industrial and systems engineering among engineering disciplines lies in the fact that it is not restricted to technological or industrial problems alone. It also covers nontechnological or non-industry-oriented problems also. The training of ISEs positions them to look at the total picture of what makes a system work best. They question themselves about the right combination of human and natural resources, technology and equipment, and information and finance. The ISEs make the system function well. They design and implement innovative processes and systems that improve quality and productivity, eliminate waste in organizations, and help them to save money or increase profitability.

Industrial and systems engineers are the bridges between management and engineering in situations where scientific methods are used heavily in making managerial decisions. The industrial and systems engineering field provides the theoretical and intellectual framework for translating designs into economic products and services, rather than the fundamental mechanics of design. Industrial and systems engineering is vital in solving today's critical and complex problems in manufacturing, distribution of goods and services, health care, utilities, transportation, entertainment, and the environment. The ISEs design and refine processes and systems to improve quality, safety, and productivity. The field provides a perfect blend of technical skills and people orientation. An industrial engineer addresses the overall system performance and productivity, responsiveness to customer needs, and the quality of the products or services produced by an enterprise. Also, they are the specialists who ensure that people can safely perform their required tasks in the workplace environment. Basically, the field deals with analyzing complex systems, formulating abstract models of these systems, and solving them with the intention of improving system performance.

The discussions under this section mainly consist of some explanations of the areas that exist for industrial and systems engineering programs in major higher institutions the world over.

HUMAN SYSTEMS

Fredrick Taylor's 1919 landmark treatise, *The Principles of Scientific Management*, discussed key methods of improving human productivity in systems (Miller et al., 2014). These techniques included selecting individuals compatible with their assigned task, tasking an appropriate number of individuals to meet time demands, providing training to the individuals, and designing work methods and implements with productivity in mind (Taylor, 1919). Taylor asserted that through improvements, which resulted in increased human productivity, the cost of manpower could be significantly reduced to improve "national efficiency." That is, the ratio of national manufacturing output to the cost of manufacture could be significantly increased. Over the intervening decades between Taylor's *Principles* and today, the domains of *manpower, personnel, training*, and *human factors* have evolved to independently address each of these respective needs. For instance, the manpower domain defines appropriate staffing levels, while the personnel domain addresses the recruitment, selection, and retention of individuals to achieve those staffing levels. The training domain ensures that each worker has the appropriate knowledge, skills, and abilities to perform their assigned tasks. And the human factors domain focuses on the design and selection of tools that effectively augment human capabilities to improve the productivity of each worker.

Taylor demonstrated, using concrete examples taken from industry (e.g., a steel mill), that selecting appropriate individuals for a task (e.g., selecting physically strong individuals for moving heavy pig iron) could allow more work to be done by fewer people. Training the individuals with the best work practices (e.g., training and incentivizing individuals to avoid nonproductive steps) also increased individual worker productivity. Likewise, improving tool selection (e.g., using larger shovels for moving light coke and smaller shovels for moving heavier coal) had a similar effect. While each of these observations was readily made in early 20th-century steel mills, an increasing amount of today's work is cognitive rather than physical in nature—and it is often performed by distributed networks of individuals working in cyberspace rather than by workers physically collocated on a factory floor. The effect of this trend is to reduce the saliency of Taylor's examples that largely focused on manual tasks (Miller et al., 2014).

It has been known for over a century that the domains of personnel, training, and human factors affect productivity and required manpower. Yet the term "human systems integration (HSI)" is relatively new, so what is HSI?

The concept of HSI emerged in the early 1980s, starting in the U.S. Army, as a more modern construct for holistically considering the domains of manpower, personnel, training, and human factors, among others. The concept has gained emphasis, both within military acquisition and the systems engineering community. The HSI concept is based on the axiom that a human-centered focus throughout the design and operation of systems will ensure that:

- Effective human-technology interfaces are incorporated in systems,
- Required levels of sustained human performance are achieved,
- Demands upon personnel resources, skills, and training are economical,
- Total system ownership costs are minimized, and
- The risk of loss or injury to personnel, equipment, and/or the environment is minimized.

HSI deals with the complexity inherent in the problem space of human performance in systems by decomposing human-related considerations into focus areas or domains (i.e., HSI analysis), which essentially form a checklist of issues that need to be considered. These domains are often aligned with specific scientific disciplines or functional areas within organizations and may vary based on the perspective and needs of individual system developers and/or owners. Equally important, the HSI concept assumes the following corollary: domains are interrelated and must be "rolled-up" and viewed holistically (i.e., HSI synthesis) to effectively understand and evaluate anticipated human performance in systems. What emerges is a view of HSI as a recursive cycle of analysis, synthesis, and evaluation, yielding HSI domain solution sets.

The decomposition of human-related considerations into domains is necessarily man-made and largely a matter of organizational convenience. Accordingly, we will not argue for the existence of an exhaustive and mutually exclusive set of HSI domains. Instead, what follows is a set of domains and their respective descriptions that have proven intuitive to an international audience:

- *Manpower/Personnel* concerns the number and types of personnel required and available to operate and maintain the system under consideration. It considers the aptitudes, experience, and other human characteristics, including body size, strength, and less tangible attributes, necessary to achieve optimum system performance. This domain also includes the necessary selection processes required for matching qualified personnel to the appropriate task, as well as tools to assess the number of individuals necessary to achieve a desired level of system performance.

- *Training* embraces the specification and evaluation of the optimum combination of instructional systems, education, and on-the-job training required to develop the knowledge, skills, and abilities (e.g., social/team-building abilities, soft skills, and competencies) needed by the available personnel to operate and maintain the system under consideration to a specified level of effectiveness under the full range of operating considerations.

- *Human Factors* are the cognitive, physical, sensory, and team dynamic abilities required to perform system-specific operational, maintenance, and support job tasks. This domain covers the comprehensive integration of human characteristics into system design, including all aspects of workstation and workspace design and system safety. The objective of this domain is to maximize user efficiency while minimizing the risk of injury to personnel and others.

- *Safety and Health* domain includes applying human factors and engineering expertise to minimize safety risks occurring as a result of the system under consideration being operated or functioning in a normal or abnormal manner. System design features should serve to minimize the risk of injury, acute or chronic illness, and/or discomfort of personnel who operate, maintain, or support the system. Likewise, design features should mitigate the risk for errors and accidents resulting from degraded job performance. Prevalent safety and health issues include noise, chemical safety, atmospheric hazards (including those associated with confined space entry and oxygen deficiency), vibration, ionizing and non-ionizing radiation, and human factors issues that can create chronic disease and discomfort such as repetitive motion injuries. Human factors stresses that create risk of chronic disease and discomfort overlap with occupational health considerations. These issues directly impact crew morale.

- *Organizational and Social* domain applies tools and techniques drawn from relevant information and behavioral science disciplines to design organizational structures and boundaries around clear organizational goals to enable people to adopt an open culture, improving sharing and trust between colleagues and coalition partners. This domain focuses on reducing the complexity of organizations. Although pertinent to all organizations, this domain is particularly germane to modern systems employing network-enabled capabilities as successful operation of these systems requires trust and confidence to be built between people in separate organizations and spatial locations who need to collaborate on a temporary basis without the opportunity to build personal relationships.

- *Other Domains* have been proposed and are worth brief mention. While the above areas have consistently been included in the HSI literature, other domains could include *Personnel Survivability, Habitability*, and the *Environment*. Personnel survivability, a military focus area, assesses designs that reduce risk of fratricide, detection, and the probability of being attacked and enable the crew to withstand man-made or natural hostile environments without aborting the mission or suffering acute or chronic illness or disability/death. Habitability addresses factors of living and working conditions that are necessary to sustain the morale, safety, health, and comfort of the user population which contribute directly to personnel effectiveness. Lastly, environmental design factors concern water, air, and land pollution and their interrelationships with system manufacturing, operation, and disposal.

HUMAN FACTORS ENGINEERING

Human factors engineering is a practical discipline dealing with the design and improvement of productivity and safety in the workplace. It concerns the relationship of manufacturing and service technologies interacting with humans. Its focus is not restricted to manufacturing alone—it extends to service systems as well. The main methodology of ergonomics involves the mutual adaptation of the components of human-machine-environment systems by means of human-centered design of machines in production systems. Ergonomics studies human perceptions, motions, workstations, machines, products, and work environments.

Today's ever-increasing concerns about humans in the technological domain make this field very appropriate. People in their everyday lives or in carrying out their work activities create many man-made products and environments for use. In many instances, the nature of these products and environments directly influences the extent to which they serve their intended human use. The discipline of human factors deals with the problems and processes that are involved in man's efforts to design these products and environments so that they optimally serve their intended use by humans. This general area of human endeavor (and its various facets) has come to be known as human factors engineering or, simply, human factors, biomechanics, engineering psychology, or ergonomics.

Operations research specifically provides the mathematical tools required by ISEs in order to carry out their task efficiently. Its aims are to optimize system performance and predict system behavior using rational decision-making and to analyze and evaluate complex conditions and systems.

This area of industrial and systems engineering deals with the application of scientific methods in decision-making, especially in the allocation of scarce human resources, money, materials, equipment, or facilities. It covers such areas as mathematical and computer modeling and information technology. It could be applied to managerial decision-making in the areas of staff and machine scheduling, vehicle routing, warehouse location, product distribution, quality control, traffic-light phasing, and police patrolling. Preventive maintenance scheduling, economic forecasting, experiment design, power plant fuel allocation, stock portfolio optimization, cost-effective environmental protection, inventory control, and university course scheduling are some of the other problems that could be addressed by employing OR.

Subjects such as mathematics and computer modeling can forecast the implications of various choices and identify the best alternatives. The OR methodology is applied to a wide range of problems in both public and private sectors. These problems often involve designing systems to operate in the most effective way. Operations research is interdisciplinary and draws heavily on mathematics. It exposes graduates in the field of industrial and systems engineering to a wide variety of opportunities in areas such as pharmaceuticals, ICT, financial consultancy services, manufacturing, research, logistics and supply-chain management, and health. These graduates are employed as technical analysts with prospects for managerial positions. Operations research adopts courses from computer science, engineering management, and other engineering programs to train students to become highly skilled in quantitative and qualitative modeling and the analysis of a wide range of systems-level decision problems. It focuses on productivity, efficiency, and quality. It also affects the creative utilization of analytical and computational skills in problem solving, while increasing the knowledge necessary to become truly competent in today's highly competitive business environment. Operations research has had a tremendous impact on almost every facet of modern life, including marketing, the oil and gas industry, the judiciary, defense, computer operations, inventory planning, the airline system, and international banking. It is a subject of beauty whose applications seem endless.

The aim of studying artificial intelligence (AI), as a system, is to understand how the human mind works, thereby fostering leading to an appreciation of the nature of intelligence and to engineer systems that exhibit intelligence. Some of the basic keys to understanding intelligence are vision, robotics, and language. Other aspects related to AI include reasoning, knowledge representation, natural language generation (NLG), genetic algorithms, and expert systems. Studies on reasoning have evolved from the following dimensions: case-based, nonmonotonic, model, qualitative, automated, spatial, temporal, and common sense. For knowledge representation, knowledge bases are used to model application domains and to facilitate access to stored information. Knowledge representation originally concentrated around protocols that were typically tuned to deal with relatively small knowledge bases but that provided powerful and highly expressive reasoning services. Natural language generation systems are computer software systems that produce texts in English and other human languages, often from nonlinguistic input data. Natural language generation systems, like most AI systems, need substantial amounts of knowledge that is difficult to acquire. In general terms, these problems were due to the complexity, novelty, and poorly understood nature of the tasks our systems attempted and were worsened by the fact that people write so differently. A genetic algorithm is a search algorithm based on the mechanics of natural selection and natural genetics. It is an iterative procedure that maintains a population of structures that are candidate solutions to specific domain challenges. During each generation, the structures in the current population are rated for their effectiveness as solutions, and on the basis of these evaluations, a new population of candidate structures is formed by using specific genetic operators such as reproduction, cross over, and mutation. An expert system is a computer software that can solve a narrowly defined set of problems using information and reasoning techniques normally associated with a human expert. It could also be viewed as a computer system that performs at or near the level of a human expert in a particular field of endeavor.

A model is a simplified representation of a real system or phenomenon. Models are abstractions revealing only the features that are relevant to the real system behavior under study. In industrial and systems engineering, virtually all areas of the disciplines have concepts that can be modeled in one form or the other. In particular, mathematical models are elements, concepts, and attributes of a real system represented by using mathematical symbols, e.g., v-5-u-1-at, A-5-r2. Models are powerful tools for predicting the behavior of a real system by changing some items in the models to detect the reaction of changes in the behavior of other variations. They provide frames of reference by which the performance of the real system can be measured. They articulate abstractions, thereby enabling us to distinguish between relevant and irrelevant features of the real system. Models are prone to manipulations more easily in a way that the real systems are often not.

In order to survive in the competitive environment, significant changes should be made in the ways of preparing organizations' design and manufacturing, selling, and servicing their goods and commodities. Manufacturers are committed to continuous improvement in product design, defect levels, and costs. This is achieved by fusing the designing, manufacturing, and marketing into a complete whole. Manufacturing system consists of two parts: its science and automation. Manufacturing science refers to investigations on the processes involved in the transformation of raw materials into finished products. This involves the traditional aspects. Traditionally, manufacturing science may refer to the techniques of work-study, inventory systems, material-requirement planning, etc. On the other hand, the automation aspect of manufacturing covers issues like e-manufacturing, Toyota Production System, the use of computer-assisted manufacturing systems (NC, CNC, and DNC), automated material handling systems, group technology, flexible manufacturing systems, and process planning and control.

Industrial and systems engineering students conduct research in the areas of manufacturing in combination with courses in finance, manufacturing processes, and personnel management. They also do research on manufacturing-design projects. This exposes the students to a manufacturing environment with activities in the design or improvement of manufacturing systems, product design, and quality.

Recent years have experienced increasing use of statistics in the industrial and systems engineering field. Industrial and systems engineers need to understand the basic statistical tools to function in a world that is becoming increasingly dependent on quantitative information. This clearly shows that the interpretation of practical and research results in industrial and systems engineering depends to a large extent on statistical methods. Statistics is used in almost every area relevant to these fields. It is utilized as a tool for evaluating economic data in "financial engineering." For this reason, ISEs are exposed to statistical reasoning early in their careers. Industrial and systems engineers also employ statistical techniques to establish quality control techniques. This involves detecting an abnormal increase in defects, which reflects equipment malfunction. The question of what, how, and when do we apply statistical techniques in practical situations and how to interpret the results are answered in the topics related to statistics.

The impact of industrial and systems engineering is complex and multi-faceted. The practitioners of data analysis in industrial and systems engineering rely on a computer as it is an important and powerful tool for collecting, recording, retrieving, analyzing simple and complex problems, as well as distributing huge information in industrial and systems engineering. It saves countless years of tedious work by the ISEs. The computer removes the necessity for men to monitor and control tedious and repetitive processes. Despite the importance of computers, their potential is so little explored that their full impact is yet to be realized. There are several powerful computer programs that can reduce the complexity of solving engineering problems.

SYSTEMS PRODUCTIVITY, EFFICIENCY, AND EFFECTIVENESS

According to Miller et al. (2014), Taylor's original focus was on "national efficiency," but to what end(s) should HSI be oriented? System designers and impending system owners by necessity must compare potential solutions to exploit opportunities and select the best solution. We previously framed HSI as seeking to maximize organizational objectives (i.e., emergent properties) for some SOI at minimum lifecycle cost. Accordingly, what is the yardstick for identifying the "best" HSI domain solution set?

To answer this question, we focus on three highly interrelated but distinct terms: productivity, efficiency, and effectiveness. *Productivity* is the rate at which goods or services are produced and is typically specified as the number of completed elements per unit of labor. *Efficiency* is the ratio of useful output to the total input in any system. Therefore, productivity might be defined as the efficiency of the human operator. *Effectiveness* refers to the ability to produce a desired effect.

We now provide an illustration of the differences among these terms. Suppose that a manufacturing system is defined to be effective if it permits a company to produce an article at a cost that allows the article to be sold at a profit. If an operator is solely responsible for production of this article and is paid based on a fixed wage, this manufacturing system will require a minimum level of productivity from the operator to be effective. Increasing the operators' productivity permits the cost of the operator to be distributed across a larger number of articles. Consequently, the cost per article associated with the operator is reduced. Restated, productivity or human efficiency can contribute to effectiveness of the manufacturing system. In traditional human-centered disciplines, such as human factors engineering, it is common to measure efficiencies such as the time required to complete a task to understand or quantify the quality of an interface.

However, productivity and efficiency are not synonymous with effectiveness. Many other factors—including factors that are both internal and external to the system—can influence the effectiveness of the system. Returning to our example, if the raw materials or market demand for the article produced by the operator is not present, the manufacturing system will not be effective regardless of the efficiency of the operator. These external influences thus render the operators' efficiency meaningless with regard to overall system effectiveness. Further, if the equipment required to process the parts outside the operators' workstation cannot match the operators' production rate, the operator's efficiency again may not have an effect upon the overall system effectiveness. Therefore, effectiveness might be influenced, but is not controlled, by productivity or operator efficiency.

Effectiveness is therefore a more valuable measure than productivity and efficiency because it assesses the degree to which a system serves its organizational purpose as well as harmonizes with other systems and its environment. Given HSI's focus on maximizing effectiveness at minimum cost—that is, maximizing the ratio of effectiveness to cost—*cost effectiveness* appears to be a naturally intuitive, unidimensional measure of merit for comparing individual HSI domain solution sets. Since money is the closest thing we currently have to a universal means of exchange, it is possible to assign costs to each HSI domain (Miller et al., 2014). Thus, cost effectiveness allows comparisons of the performance of HSI domain solution sets in terms of the desired system emergent properties while also capturing the input parameter of the summation of the respective domain-related costs.

While cost effectiveness is a useful measure, it may not always result in universally better systems in terms of the owning organization. For example, depicting a SOI and three sibling systems, each consisting of a personnel subsystem and a technological subsystem. The SOI and sibling systems, in turn, are components of a larger containing system (i.e., the parent organization). If we solely focus on the SOI, we would seek the joint optimization of the personnel and technological subsystems within the SOI, thereby maximizing its cost effectiveness and contributing positively to the containing system's objectives. Now, let us assume that the SOI and its sibling systems must share a common personnel resource pool. It is possible in maximizing the cost effectiveness of the SOI to have unintended downstream effects on the personnel subsystems of the sibling systems. These downstream effects could result in decreased effectiveness of the sibling systems. In aggregate, maximizing the cost effectiveness of the SOI may actually result in a net negative contribution toward achieving the containing system's objective! Such a scenario illustrates the need to also consider the "Net Contribution" of an HSI domain solution set.

Timing is everything, or so the saying goes. The greatest value is obtained by the early involvement of HSI specialists. Between 70% and 90% of life cycle costs are already locked in by the end of conceptual system design. Given that 80% of life cycle costs are HSI-related and 40–60% is attributable to the manpower, personnel, and training domains alone, the longer one waits to begin addressing HSI, the more negative impact will be shown on total life cycle costs. Additionally, since human performance contributes significantly to system effectiveness, the only question is whether HSI will be paid for most affordably in advance, or at much greater expense after a mature system design reveals significant problems. The earlier an HSI investment can be made, the greater will be its return. However, it is important to remember that there will be benefits of incorporating HSI at any point in the design maturity, as long as it precedes the final design.

All the aforementioned characteristics of the world of systems point to the need for guiding models for understanding, managing, and advancing integrated systems of people, processes, and tools. This is one of the basic foundations for the structural framework of DEJI Systems Model. In many human endeavors, things are often naturally "unstructured." DEJI Systems Model can impart structure in many systems, whether human or digital.

REFERENCES

Oke, S. A. (2014), "An Overview of Industrial and Systems Engineering," in Badiru, A. B., editor, *Handbook of Industrial and Systems Engineering*, Second Edition, CRC Press, Boca Raton, FL (pp. 185–196).

Miller, Michael E., John Colombi, and Anthony Tvaryanas (2014), "Human Systems Integration," in Badiru, A. B., editor, *Handbook of Industrial and Systems Engineering*, Second Edition, CRC Press, Boca Raton, FL (pp. 197–216).

Taylor, Frederick W. (1919), *The Principles of Scientific Management*. Harper and Brothers Publishers, New York.

2 Overview of DEJI Systems Model®

We must temper idealism with the reality of integration.

—Adedeji Badiru

INTRODUCTION

This chapter presents a concise overview or executive summary of the elements and essence of the trademarked DEJI Systems Model for design, evaluation, justification, and integration. More explicit details of each of the four elements of the model are presented in the chapters that follow. Where some paragraphs in this chapter appear to be repeated in later chapters, it is by design (for emphasis), rather than by duplicative neglect.

Ideas (designs, concepts, proposals, etc.) that are good in principle often don't integrate well into the realities of a sustainable implementation. Therein lies the case for systems integration as advocated by the DEJI Systems Model. As an author, my love of systems integration is predicated on the belief that you can fight to get what you want and get hurt in the fighting process or you can systematically negotiate and win without shedding any tears, limited blood, or excessive sweat. It is in systems integration that we actually get things done. In a systems context, integration is everything. Several practical examples exist to ginger our interest in systems integration. One common (and relatable) example is in traffic behavior of drivers. Integration implies monitoring, adapting to, and reacting to the prevailing scenario in interstate road traffic. In this case, integration calls for observing and enmeshing into the ongoing traffic with respect to flow and speed. If everyone drives with a sense of traffic systems integration, there will be fewer accidents on the road. In the full DEJI systems framework, a driver would observe and design (i.e., formulate) driving actions, evaluate the rationale and potential consequences of the "designed" actions, justify (mentally) why the said actions are necessary and pertinent for the current traffic scenario, and then integrate the actions to the driving condition.

Integration is the basis for the success of any system. This is particularly critical for system of systems (SoS) applications where there may be many moving parts. This book presents a practical framework for applying the trademarked DEJI Systems Model® for SoS, with respect to systems design, evaluation, justification, and integration. Case examples will be presented on recent applications of the model and its efficacy in diverse problem scenarios. The DEJI Systems Model has been reported to be particularly effective for systems thinking in project planning, project evaluation, and project control and other human pursuits.

DOI: 10.1201/9781003175797-2

A familiar example is the January 2022 announcement of the retirement of U.S. Supreme Court Justice Stephen Breyer. Justice Breyer was known as a legal pragmatist, who worked to make the law work in consonance with how the society lives and works. That is a good spirit of aligning and integrating the law with people's lives and expectations. When a system is aligned with reality, the probability of success increases exponentially.

Badiru (2012) first formally introduced the trademarked DEJI Systems Model® as a structured process for accomplishing systems design, evaluation, justification, and integration in product development. The model has since been adopted and applied to other application areas, such as quality management (Badiru, 2014) and engineering curriculum integration (Badiru and Racz, 2018). The premise of the model is that integration across a system is the overriding requirement for a successful system of systems (SoS).

A system is represented as consisting of multiple parts, all working together for a common purpose or goal. Systems can be small or large, simple or complex. Small devices can also be considered systems. Systems have inputs, processes, and outputs. Systems are usually explained using a model for visual clarification of inputs, processes, and outputs. A model helps illustrate the major elements and their relationships. Figure 2.1 illustrates the basic model structure of a system.

In the ICOM chart, processes are the intervening mechanisms that convert inputs to outputs. This implies that processes represent the interactions that occur between elements in a system. In the human terms in the system, these may be viewed as the human facilitators and mediators in the system. When interrelated structures of input-process-output systems are stringed together, we have a system of systems that must be integrated structurally. Therein lies the efficacy of the integration component of the DEJI Systems Model.

Systems engineering is the application of engineering tools and techniques to the solutions of multi-faceted problems through a systematic collection and integration of parts of the problem with respect to the lifecycle of the problem. It is the branch of engineering concerned with the development, implementation, and use of large or complex data sets across diverse domains. It focuses on specific

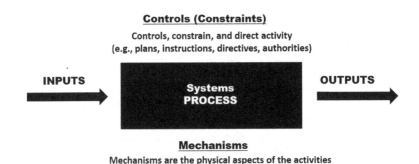

FIGURE 2.1   ICOM systems input, process, and output structure.

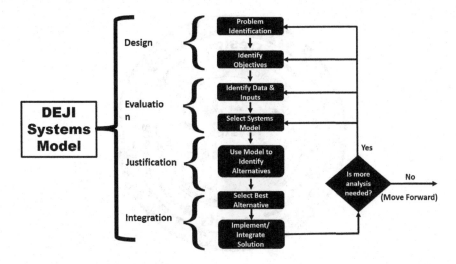

FIGURE 2.2   Flowchart of DEJI Systems Model flow framework.

goals of a system considering the specifications, prevailing constraints, expected services, possible behaviors, and structure of the system. It also involves a consideration of the activities required to ensure that the system's performance matches specified goals. Systems engineering addresses the integration of tools, people, and processes required to achieve a cost-effective and timely operation of the system. Some of the features of this book include solutions to multi-faceted problems; a holistic view of a problem domain; applications to both small and large problems; decomposition of complex problems into smaller manageable chunks; direct considerations for the pertinent constraints that exist in the problem domain; systematic linking of inputs to goals and outputs; explicit treatment of the integration of tools, people, and processes; and a compilation of existing systems engineering models. A typical decision support model is a representation of a system, which can be used to answer questions about the system. While systems engineering models facilitate decisions, they are not typically the conventional decision support systems. The end result of using a systems engineering approach is to integrate a solution into the normal organizational process. For that reason, the DEJI Systems Model is desired for its structured framework of design, evaluation, justification, and integration. The flowchart for this framework is shown in Figure 2.2.

SYSTEMS ENGINEERING COMPETENCY FRAMEWORK

The DEJI Systems Model is operationally aligned with the ideals and expectations of the INCOSE Systems Engineering Competency Framework (INCOSE SECF) developed by the International Council on Systems Engineering (INCOSE). INCOSE SECF represents a worldview of five competency groupings with 36 competencies for systems implementations. The integration requirement of the

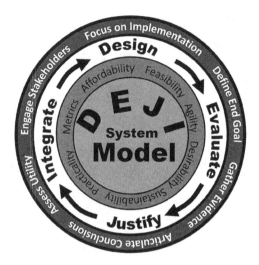

FIGURE 2.3   Trademarked logo of DEJI Systems Model®.

DEJI Systems Model directly focuses on what is required to achieve successful systems implementations. Figure 2.3 illustrates the wheel of the alignment of the DEJI Systems Model to the scope of practical systems implementations in business, industry, and other enterprises.

SYSTEM DESIGN

In as much as "design" is the origin of any system, the DEJI Systems Model starts with a structural articulation of what design embraces. In general, design can encompass a variety of pursuits ranging from physical product design to the design of conceptual framework for process improvement. In each generic frame, the systems value model (SVM) approach is effective in assessing the properties, expected characteristics, and desired qualities of a system. SVM provides an analytical decision aid for comparing system alternatives. In this case, value is represented as a p-dimensional vector:

$$V = f\left(A_1, A_2, \ldots, A_p\right),$$

where $A = (A_1, \ldots, A_n)$ is a vector of quantitative measures of tangible and intangible attributes of the system. Examples of process attributes are quality, throughput, capability, productivity, cost, and schedule. Attributes are considered to be a combined function of factors, x_1 expressed as:

$$A_k\left(x_1, x_2, \ldots, x_{m_k}\right) = \sum_{i=1}^{m_k} f_i\left(x_i\right)$$

where $\{x_i\}$ = set of m factors associated with attribute $A_k(k = 1, 2, \ldots, p)$ and f_i = contribution function of factor x_i to attribute A_k. Examples of factors include reliability, flexibility, user acceptance, capacity utilization, safety, and design functionality. In fact, factors can be expanded to cover many of the common "ilities" of a system, such as affordability, practicality, desirability, configurability, modularity, desirability, maintainability, testability, reachability, and agility. Factors are themselves considered to be composed of indicators, v_i expressed as

$$x_i\left(v_1, v_2, \ldots, v_n\right) = \sum_{j=1}^{n} z_i\left(v_i\right)$$

where $\{v_j\}$ = set of n indicators associated with factor $x_i(i = 1, 2, \ldots, m)$ and z_j = scaling function for each indicator variable v_j. Examples of indicators are project responsiveness, lead time, learning curve, and work rejects. By combining the above definitions, a composite measure of the value of a process can be modeled as:

$$V = f\left(A_1, A_2, \ldots, A_p\right)$$

$$= f\left\{\left[\sum_{i=1}^{m_1} f_i\left(\sum_{j=1}^{n} z_j\left(v_j\right)\right)\right]_1, \left[\sum_{i=1}^{m_2} f_i\left(\sum_{j=1}^{n} z_j\left(v_j\right)\right)\right]_2, \ldots, \left[\sum_{i=1}^{m_k} f_i\left(\sum_{j=1}^{n} z_j\left(v_j\right)\right)\right]_p\right\}$$

where m and n may assume different values for each attribute. A subjective measure to indicate the utility of the decision-maker may be included in the model by using an attribute weighting factor, w_i, to obtain a weighted PV:

$$PV_w = f\left(w_1 A_1, w_2 A_2, \ldots, w_p A_p\right),$$

where

$$\sum_{k=1}^{p} w_k = 1, \quad \left(0 \le w_k \le 1\right).$$

With this modeling approach, a set of design options can be compared on the basis of a set of attributes and factors, both quantitative and qualitative. To illustrate the model above, suppose three IT options are to be evaluated based on four attribute elements: *capability, suitability, performance,* and *productivity.* For this example, based on the equations, the value vector is defined as:

$$V = f\left(\text{capability, suitability, performance, productivity}\right)$$

Capability: The term "capability" refers to the ability of IT equipment to satisfy multiple requirements. For example, a certain piece of IT equipment may only provide computational service. A different piece of equipment may be capable of generating reports in addition to computational analysis, thus increasing the service variety that can be obtained. In the analysis, the levels of increase in service variety from the three competing equipment types are 38%, 40%, and 33% respectively. *Suitability:* "Suitability" refers to the appropriateness of the IT equipment for current operations. For example, the respective percentages of operating scope for which the three options are suitable for are 12%, 30%, and 53%. *Performance:* "Performance," in this context, refers to the ability of the IT equipment to satisfy schedule and cost requirements. In the example, the three options can, respectively, satisfy requirements on 18%, 28%, and 52% of the typical set of jobs. *Productivity:* "Productivity" can be measured by an assessment of the performance of the proposed IT equipment to meet workload requirements in relation to the existing equipment. For the example, the three options, respectively, show normalized increases of 0.02, −1.0, and −1.1 on a uniform scale of productivity measurement. Option C is the best "value" alternative in terms of suitability and performance. Option B shows the best capability measure, but its productivity is too low to justify the needed investment. Option A offers the best productivity, but its suitability measure is low. The analytical process can incorporate a lower control limit into the quantitative assessment such that any option providing value below that point will not be acceptable. Similarly, a minimum value target can be incorporated into the graphical plot such that each option is expected to exceed the target point on the value scale. The relative weights used in many justification methodologies are based on subjective propositions of decision-makers. Some of those subjective weights can be enhanced by the incorporation of utility models. For example, the weights could be obtained from utility functions. There is a risk of spending too much time maximizing inputs at "point-of-sale" levels with little time defining and refining outputs at the "wholesale" systems level.

A systems view of operations is essential for every organization. Without a systems view, we cannot be sure we are pursuing the right outputs that can be *integrated* into the prevailing operating environment. Thus, the DEJI Systems Model allows for a multi-dimensional analysis of any endeavor, considering many of the typical "ilities" related to system of systems.

SYSTEM EVALUATION

Evaluation is the second stage of the structural application of the DEJI Systems Model. The evaluation of a system pursuit can range over a variety of rubrics related to organizational performance metrics. In many cases, the basic requirement of economic evaluation may be needed beyond the technical sphere provided in the design phase of the DEJI Systems Model. Other basis of the evaluation of a system may involve materials availability, quality expectation, supply chain reliability, skilled workforce, and so on. For example, an evaluation of a skilled workforce to run a well-designed system may be needed to ensure that the system can, indeed, be operated. In this case, operations research techniques of optimizing work assignment could be useful. For example, a resource-assignment algorithm can be used to enhance the quality of resource-allocation decisions. Suppose there are n tasks which must be performed by n workers. The cost of worker i performing task j is c_{ij}. It is desired to assign workers to the tasks in a fashion that minimizes the cost of completing the tasks. This problem scenario is referred to as the assignment problem. The technique for finding the optimal solution to the problem is called the assignment method. The assignment method is an iterative procedure that arrives at the optimal solution by improving on a trial solution at each stage of the procedure. Although the assignment method is cost-based, task duration can be incorporated into the modeling in terms of time-cost relationships for a more robust evaluation, based on the organization's interest. If the objective is to minimize the cost of the system implementation, the formulation of the assignment problem can be as shown below:

Let

$x_{ij} = 1$ if worker i is assigned to task j, $j = 1, 2, \ldots, n$,

$x_{ij} = 0$ if worker i is not assigned to task j,

c_{ij} = cost of worker i performing task j.

$$\text{Minimize: } z = \sum_{i=1}^{n} \sum_{j=1}^{n} c_{ij} x_{ij}$$

$$\text{Subject to: } \sum_{j=1}^{n} x_{ij} = 1, \quad i = 1, 2, \ldots, n$$

$$\sum_{i=1}^{n} x_{ij} = 1, \quad j = 1, 2, \ldots, n$$

$$x_{ij} \geq 0, \ i, j = 1, 2, \ldots, n$$

The above formulation uses the non-negativity constraint, $x_{ij} \geq 0$, instead of the integer constraint, $x_{ij} = 0$ or 1. However, the solution of the model will still be integer-valued. Hence, the assignment problem is a special case of the common transportation problem in operations research, with the number of sources (m) = number of targets (n), $S_i = 1$ (supplies), and $D_i = 1$ (demands). The basic requirements of an assignment problem are summarized below:

1. There must be two or more tasks to be completed;
2. There must be two or more resources that can be assigned to the tasks;
3. The cost of using any of the resources to perform any of the tasks must be known;
4. Each resource is to be assigned to one and only one task.

If the number of tasks to be performed is greater than the number of workers available, we will need to add *dummy workers* to balance the problem. Similarly, if the number of workers is greater than the number of tasks, we will need to add *dummy tasks* to balance the problem. If there is no problem of overlapping, a worker's time may be split into segments so that a worker can be assigned to more than one task. In this case, each segment of the worker's time will be modeled as a separate resource in the assignment problem. Thus, the assignment problem can be extended to consider partial allocation of resource units to multiple tasks. The example presented here is just to illustrate what the evaluation stage of the DEJI Systems Model may entail. Other evaluation parameters and quantitative solution approaches are available for users to consider, based on internal interests and tools available within the organization.

SYSTEM JUSTIFICATION

Justification is the third stage of the application of DEJI Systems Model. In this case, justification goes beyond the typical economic justification in project planning and control. While economic feasibility may be included in the evaluation stage of DEJI Systems Model, it is not the necessary basis of a systems justification. Justification, in the context of the DEJI System Model, is often more qualitative and conceptual than quantitative. Not all systems that are well-designed and favorably evaluated are justified for implementation. Questions related to systems justification may include the following:

- Desirability of the system for the operating environment.
- Acceptability of the system by those who will be charged to run the system.
- Safety protocols related to operating the system.
- Regulatory oversight for operating the system.
- Compliance with industry standards for operating the system.
- Organizational philosophy about system expansion.
- Investment potential for actualizing the system.
- Sustainability potential for the new system.

The beauty of the DEJI Systems Model is that it explicitly requires or "forces" an organization to justify why a new is required. In many practical applications of systems modeling, some discoveries and revelations have pointed to discouraging realities that were not previously recognized or highlighted during the system realization stage (Valerdi 2014).

SYSTEMS INTEGRATION

Integration is the fourth and last stage of the application of the DEJI Systems Model, and deferentially the most critical part of any systems implementation. A system that is not appropriately anchored to the prevailing operating environment may be doomed to failure. This fact is often the reason behind many system failures that are seen in practice. The DEJI System Model explicitly requires that any system of interest be *integrated* into its point and manner of use. Integration can be affirmed through qualitative or quantitative mechanisms. The following are some of the pertinent questions for integration:

- Is the new system aligned with the existing operating framework of the organization?
- Is the new system in line with the prevailing mission of the organization?
- Can this system really do what it is designed to do?
- Is the prevailing work environment able to assimilate an implementation of the new system?
- What are the practical implementation requirements for adopting the new system?
- What operating constraints exist to successfully implement the new system?

With probing questions such as the above, DEJI System Model can positively impact how new systems are brought online in the prevailing operating environment, considering the existing tools, techniques, processes, and workforce with the organization. Without being integrated, a system will be in isolation and it may be worthless. We must integrate all the elements of a system on the basis of alignment of functional goals. In the case of a quantitative process, the overlap of systems for integration purposes can be viewed, conceptually, as projection integrals by considering areas bounded by the common elements of subsystems. This approach is useful for System of Systems applications, where each independent system is a subsystem of the larger SoS. Quantitative metrics can be applied at this stage for effective assessment of the composite system. Trade-off analysis is essential in system integration. Pertinent questions include the following:

- What levels of trade-offs on the level of integration are tolerable?
- What is the incremental cost of pursuing higher integration?
- What is the marginal value of higher integration?
- What is the adverse impact of a failed integration attempt?

What is the value of integration of system characteristics over time? In this respect, an integral of the form below may be suitable for a mathematical exposition:

$$I = \int_{t_1}^{t_2} f(q)\,dq,$$

where I = integrated value of quality, $f(q)$ = functional definition of quality, t_1 = initial time, and t_2 = final time within the planning horizon. An illustrative example is provided in Figure 2.4 for the case of a geometric alignment of the hypothetical physical parts of SoS.

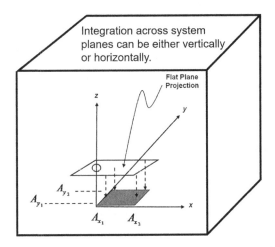

FIGURE 2.4   Systems integration geometric surfaces.

Presented below are guidelines and important questions relevant for system integration.

- What are the unique characteristics of each component in the integrated system?
- How do the characteristics complement one another?
- What physical interfaces exist among the components?
- What data and information interfaces exist among the components?
- What ideological differences exist among the components?
- What are the data flow requirements for the components?
- What internal and external factors are expected to influence the integrated system?
- What are the relative priorities assigned to each component of the integrated system?
- What are the strengths and weaknesses of the integrated system?
- What resources are needed to keep the integrated system operating satisfactorily?
- Which organizational unit has primary responsibility for the integrated system?

Other quantitative alignment and integration methodologies are available in the literature for consideration in this last stage of the DEJI System Model.

CONCLUSION

Outputs of SoS follow an integrative process that must be evaluated on a stage-by-stage approach. This requires research, education, and implementation strategies that consider several pertinent factors. This paper suggests the DEJI System

Model, which has been used successfully for product development applications, as a viable methodology for system design, system evaluation, system justification, and system integration. The integrative approach of the DEJI System Model will facilitate a better alignment of technology with future development and needs. The stages of the model require research for each new product with respect to design, evaluation, justification, and integration. Existing analytical tools and techniques as well as other systems engineering models can be used at each stage of the model. Thus, a hybrid systems modeling is possible. It is anticipated that this paper will spark the interest of researchers and practitioners to apply this tool in new system development initiatives. For example, in a contested political environment, an engineering systems approach can be applied to evaluate and justify a politically driven national proposal, with a specific template for integration into the expectations of the local population. This can help mitigate the common realizations of failed political promises. Such an application will expand the applicability of the DEJI System Model to unconventional platforms of national debates related to economic development.

REFERENCES

Badiru, Adedeji B. (2012), 'Application of the DEJI Model for Aerospace Product Integration', *Journal of Aviation and Aerospace Perspectives (JAAP)*, Vol. 2, No. 2, pp. 20–34, Fall 2012.

Badiru, Adedeji B. (2014), 'Quality Insights: The DEJI Model for Quality Design, Evaluation, Justification, and Integration', *International Journal of Quality Engineering and Technology*, Vol. 4, No. 4, pp. 369–378.

Badiru, Adedeji B. and Ann Racz (2018), 'Integrating Systems Thinking in Interdisciplinary Education Programs: A Systems Integration Approach,' *Proceedings of the Annual Conference of the American Society for Engineering Education (ASEE)*, Salt Lake City, UT, June 2018.

Valerdi, R. (2014), "Systems engineering cost estimation with a parametric model," Chapter in A. B. Badiru, editor, *Handbook of Industrial and Systems Engineering*, CRC Press, Boca Raton, FL.

3 Design in DEJI Systems Model®

INTRODUCTION

Design is at the root of all enterprise pursuits. It goes beyond what we typically recognize as the physical design of a product. Design can embody the design and articulation of physical products, services, and results (organizational expectations). Design, in the context of the DEJI Systems Model®, can cover various organizational pursuits that may range from the design of physical products, process design, enterprise framework, concept blueprint, and other tactical or strategic planning. The origin of the DEJI Systems Model (Badiru, 2012) goes back to a design integration research project funded by the National Science Foundation (NSF) at the University of Oklahoma in 1994. Design, in its diverse ramifications (physical, conceptual, quantitative, or qualitative), is about making a decision about options for achieving the desired characteristics of the product of interest.

The traditional design environment involves segregated cubicles of designers who are confined to their own worlds of design ideas. This approach to design worked in the distant past because consumers were less sophisticated and the market was defined more by whatever design was available to the consumers. We recall the famous 1909 quote by Henry Ford: "Any customer can have a car painted any color that he wants so long as it is black."

In the present day of globalized markets, designs must be more responsive to the changing environment of the market. More of the design and manufacturing resources available within an organization must be utilized in an integrated fashion in order to create competitive designs. A competitive design will need to be not only effective but also timely. Timeliness of design requires communication, cooperation, and coordination (Badiru, 2008) of design and manufacturing functions. Design information must be shared in a timely, accurate, efficient, and cost-effective manner.

The NSF-funded research addresses the development of a performance measurement methodology for design process integration in manufacturing systems. The methodology uses a hybrid model of state-space representation and expert system. The design process is based on the Pahl-and-Beitz systematic approach to design. The Triple C model of project management (Badiru, 2008) is used to facilitate the communication, cooperation, and coordination required for design integration. Many conceptual integration approaches are available in the literature. But most of these existing approaches lack a quantitative performance measure to drive integration efforts. The research focuses on the design process rather than just the physical design itself. For the purpose of the research, we define the design process as presented below:

> The design process is the collection of all the activities and functions that support a design effort. These include the qualitative and quantitative properties of the physical design.

To achieve an effective and timely design process, the research uses a hybrid model of state-space representation and expert system to quantify the state and progress of the design process. The research involves an approach that models the state-to-state transformation of design process within a concurrent engineering framework. Our research, for the first time, presents a quantifiable and measurable approach that is usable within the software, hardware, and human constraints of the design process. We recall again that design, in the context of the DEJI Systems Model, encompasses more than just the design of a physical product. That is the essence of how the DEJI Systems Model evolved from the NSF research on design integration. The integrative framework for the proposed model is represented by linking a quantitative state-space model, design communication/coordination, and the expert system model into a real-time monitoring tool to generate a quantitative performance measure. The state-space representation model is the primary vehicle for linking the various components in the design process.

HUMAN ASPECTS OF DESIGN

Peacock (2009, 2019) presents comprehensive coverages of the multi-dimensional aspects of design, particularly where the human elements are involved. Peacock (2009) presents the grammar of design, as depicted below and in Figure 3.1:

- A **System** is a collection of elements such as machines, people, and operational processes.
  - Systems are specified by nouns and quantified by adjectives.
  - System designs can be verified.
- A **Process** is the interaction of two or more systems with a defined purpose.
  - Processes are described by verbs and quantified by adverbs.
  - Process requirements can be validated.
- The **Context or Conditions** of a process will use other linguistic conventions such as conjunctions and prepositions.

Verification and validation across the design landscape will encompass the following considerations:

- Design specifications
  - Adjectives, which can be verified
- Performance requirements
  - Adverbs, which can be validated

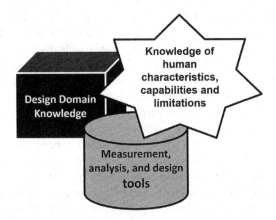

FIGURE 3.1   Generic design requirements for human in the loop.

In any design, we must pay attention to both the small and big things. The big picture is particularly of "big importance." The following age-old poem is of relevance here:

> For the want of a nail the shoe was lost;
> For the want of a shoe the horse was lost;
> For the want of a horse the rider was lost;
> For the want of a rider the battle was lost;
> For the want of a battle the kingdom was lost;
> And all for the want of a horseshoe nail

This poem has its origins in the 14th century, and Benjamin Franklin wrote a variation on the theme around the time of the war of independence. It is central to the understanding of design as it relates to the details of human in the loop of any system, centered on looking at the big picture. The nail is a metaphor that represents all the critical components of an integrated system as they contribute to a purposeful process. Single point failures abound and most of them involve "Human the User."

A DESIGN CASE STUDY

In a case study presented by Peacock (2019), the General Motors ACCESS car program involved studies by many academic institutions, a series of customer clinics with a wide variety of vehicles and the design and evaluation of some 200 features. The features were classified into physical, informational, and psycho-socio-economic. The entry-egress features of interest were related to the design of door openings, including step over and head clearance, seat height and profile, obstructions and the availability of support devices, such as grab handles, seat backs, instrument panels, and steering wheels. Similar consideration was given to storage, for large and small articles, both within the vehicle interior and the trunk. Seat design and seat comfort faced the inevitable discussion of the differences between "comfort" and "discomfort"; on balance, however, the older drivers preferred and performed better with firmer seats with less contouring. Access to restraint systems was also a concern with the placement of the seat belts and the fastening mechanisms. Also, the particular fragility and vulnerability of older people to airbags was and remains a concern.

At the informational level, there was polarization between the demand of many, usually younger customers, for features of all varieties to enhance their driving and peripheral experience, and the requirement of older people for simplicity. Much of this debate revolved around access to a hierarchy of features and interface designs. The vehicle system information displays followed the classical sequence of the following:

• Tell the driver about the existence of a deviation from normal.
• Indicate the seriousness of the deviation.
• Indicate the source.
• Communicate what should be done to resolve the problem.
• Indicate the urgency of the problem.

The vehicle operation information display evidence agreed on the visual needs of older people for size and contrast but bounced back and forth between the merits of status ("idiot lights"), analog, digital, and representational displays. The special needs of older drivers were addressed by the development of an emergency communication system that at the press of a (large) button connected the driver with assistance for medical, vehicle, navigational, and security assistance. This system was a prototype for the commonly available driver communication facilities that involve GPS (Global Positioning System) and agent-broker systems. The agent-broker concept was also applied to the psycho-social-economic aspects of transportation. Ownership of a vehicle has many challenges—purchase, licensing, insurance, driving, maintenance, repair, disposal, etc. The provision of mass, small group, and personal transport for older adults requires a high-level systems approach, that started with analysis of types of trips taken. One outcome of this analysis resulted in the development of a neighborhood car concept that required attention to vehicle design, organizational access, and journey management. The design implementation of this analysis resulted in the confinement of small electric or hybrid vehicles to largely self-contained neighborhoods, isolated from the 18-wheelers, frantic commuters, and distracted teenagers. The more recent development of widely accessible ride-share systems (Uber, Lyft, etc.) has developed to address the various requirements of many, including older people. This is a good demonstration of design integration with the customer base. An example of General Motors' customer clinics for system desirability is presented by Badiru and Neal (2013).

INTEGRATIVE DESIGN PROCESS

Designing is the intellectual attempt to meet certain demands in the best possible way. It is an activity that impinges on nearly every sphere of human life and organizational strategies. It relies on the discoveries and laws of science, and human factors and creates the conditions for applying process laws to the manufacture of useful products. In psychological respects, designing is a creative activity that calls for a sound grounding in mathematics, physics, chemistry, mechanics, thermodynamics, hydrodynamics, electrical engineering, production engineering, materials technology, and design theory, together with practical knowledge and experience in specialist fields. In this regard, design is all encompassing. Everything we do starts with some sort of a design framework, and that is the essence of the efficacy of the DEJI Systems Model.

In systematic respects, designing is the optimization of objectives within partly conflicting constraints. Requirements change with time, so a particular solution can only be optimized in a particular set of circumstances. In organizational respects, designing plays an essential part in the manufacture and processing of raw materials and products. It calls for close collaboration with workers in many other spheres. Thus, to collect all the information it needs, the designer must establish close links with salesmen, buyers, cost accountants, estimators, planners, production engineers, materials specialists, researchers, test engineers, and standards engineers. In essence, design facilitates teamwork across occupational specialties.

An essential part of the design methodology involves step-by-step analysis and synthesis. In this method, we proceed from the qualitative to the quantitative, each new step being more concrete than the last. So, design is progressive from the raw material stage to the finished product. Conversion of information at each step not only provides data for the next step but also sheds new light on the previous step.

In manufacturing systems, computer-aided design (CAD) and computer-aided manufacturing (CAM) are subsets of the design and manufacturing sub-processes of the general design process. CAD tools are the intersection of three sets consisting of the following:

- Geometric modeling
- Computer graphics
- Design tools

In a general system for the application of DEJI Systems Model, design goes beyond the above three elements. It is more flexible and more accommodating in allowing both conceptual and physical elements of how an organization does its business. Design tools can include analysis codes, such as spreadsheets, equations solvers, parametric/variational, stress and strain, kinematic and dynamic, and fluid and thermal, heuristic procedures, such as expert systems and design practices. CAM tools can be defined as the intersection of three sets:

- CAD tools
- Networking concepts
- Manufacturing tools

Manufacturing tools include process planning, NC (numerical control) programs (manual or computer based), inspection, assembly, and packaging. Process planning is the backbone of the manufacturing process since it determines the most efficient sequence to produce the product. The outcome of process planning is the production path, tools procurement, material order, and machine programming. The design process encompasses various components that must be integrated. This integration is not possible without a quantifiable performance measure to indicate the state of each component.

DESIGN COMMUNICATION, COOPERATION, AND COORDINATION

As in general project management, design management is the process of managing, allocating, and timing resources to achieve a desired design goal in an efficient and expedient manner. The objectives that constitute the goal are in terms of time, costs, and performance. Communication, cooperation, and coordination using the Triple C model of project management (Badiru, 2008) are essential in the design environment. Since people will be the facilitators of the design process, the project management approach, embedded within DEJI Systems Model, explicitly accounts for the human elements by using the Triple C model. The Triple C model is an effective project management approach that states that project management can be enhanced by implementing it within the integrated functions of communication, cooperation, and coordination. The model facilitates a systematic approach to planning, organizing, scheduling, and control in any organizational strategy. All of the above requirements for design integration will be possible only if a reliable and quantifiable measure of the state of design is available to facilitate human-to-human, human-to-machine, and machine-to-machine communication. Such a measure is provided by state-space representation.

STATE-SPACE REPRESENTATION

A state is a set of conditions that describes the design process at a specified point in time. A formal definition of state in the context of the proposed research is presented below:

> The state of a design refers to a performance characteristic of the design which relates input to output such that knowledge of the input time function for $t \leq t_0$ and the state at time $t = t_0$ determines the expected output for $t \geq t_0$.

A design state space is the set of all possible states of the design process. State-space representation can solve design problems by moving from an initial state to another state, and eventually to a goal state. The movement from state to state is achieved by the means of design actions. This is akin to how a team in an organization moves forward with goals and objectives by sequentially linking individual and collective actions. A goal is a description of an intended state that has not yet been achieved. The process of solving a design problem involves finding a sequence of design actions that represents a solution path from the initial state to the goal state.

A decision system (expert or heuristic), linked to a state-space model, can improve the management of the general design process. A state-space model consists of state variables that describe the prevailing condition of the design process. The state variables are related to the design inputs by mathematical relationships. Examples of potential design state variables include product functionality, quality, cost, due date, resource, skill, and productivity level. For a process described by a system of differential equations, the state-space representation is of the form:

$$\dot{z} = f\big(z(t), x(t)\big)$$
$$y(t) = g\big(z(t), x(t)\big)$$

where f and g are vector-valued functions. For linear systems, the representation is:

$$\dot{z}' = \mathbf{A}z(t) + \mathbf{B}x(t)$$
$$y(t)' = \mathbf{C}z(t) + \mathbf{D}x(t)$$

where $z(t)$, $x(t)$, and $y(t)$ are vectors and $\mathbf{A}$, $\mathbf{B}$, $\mathbf{C}$, and $\mathbf{D}$ are matrices. The variable y is the output vector, while the variable x denotes the inputs. The state vector $z(t)$ is an intermediate vector relating $x(t)$ to $y(t)$. The state-space representation of a discrete-time linear dynamic design system is represented as:

$$z(t+1) = \mathbf{A}z(t) + \mathbf{B}x(t)$$
$$y(t) = \mathbf{C}z(t) + \mathbf{D}x(t)$$

In generic terms, a design is transformed from one state to another by a driving function that produces a transitional equation given by:

$$S_s = f\big(x|S_p\big) + \varepsilon$$

where S_s is the subsequent state, x is the state variable, S_p is the preceding state, and ε is the error component. The function f is composed of a given action (or a set of actions) applied to the design process. Each intermediate state may represent a significant milestone in the process. Thus, a descriptive state-space model facilitates an analysis of what actions to apply in order to achieve the next desired design state. Although a technical application of the state-space modeling approach can be very mathematical, the usage in a real human environment can be primarily heuristic. The design phase of DEJI Systems Model does not have to be mathematical. It just has to prompt humans (actors) to be cognizant of their actions and the consequences the actions pose. It is through this structured approach that organizational functions can move forward sequentially from one point to another toward the intended end goal.

DESIGN STATE TRANSFORMATION PATHS

Design objectives are achieved by state-to-state transformation of successive design phases or action paths. Figure 3.2 illustrates this conceptual framework to show the transformation of a design element (or process action) from one stage to another through the application of actions. This simple representation can be expanded to cover several components within the design framework. The dashed lines represent additional sequence of actions not shown in the diagram. The hierarchical linking of design elements provides an expanded transformation structure.

The design state can be expanded in accordance with implicit design requirements. These requirements might include grouping of design elements, precedence linking (both technical and procedural), required communication links, and

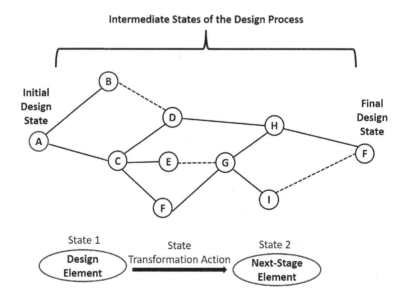

FIGURE 3.2 Transformation of design state paths.

reporting requirements. The actions to be taken at each state depend on the prevailing design conditions. The natures of the subsequent alternate states depend on what actions are implemented. Sometimes, there are multiple paths that can lead to the desired end result. At other times, there exists only one unique path to the desired objective. In conventional practice, the characteristics of the future states can only be recognized after the fact. This makes it impossible to develop adaptive plans. In the framework of DEJI Systems Model, adaptive plans can be achieved because the events occurring within and outside the design state boundaries can be taken into account. Thus, environmental factors can be considered, recognized, or acknowledged in the overall design process. This general framework permits the inclusion of scientific principles, managerial policies, approved procedures, technical information, designer's creativity, performance expectation, time limitation, resource constraints, human factors, ergonomics, and product specifications within the composite design environment.

REAL-TIME STATE MONITORING

In general, if we describe a product by P state variables s_i, then the composite state of the product at any given time can be represented by a vector S containing P elements. That is,

$$S = \{s_1, s_2, \ldots, s_P\}$$

The components of the state vector could represent either quantitative or qualitative variables (e.g., cost, energy, color, time). We can visualize every state vector as a point in the state space of the product. The representation is unique since every state vector corresponds to one and only one point in the state space. Suppose we have a set of actions (transformation agents) that we can apply to the product information so as to change it from one state to another within the project state space. The transformation will change a state vector into another state vector. A transformation may be a change in raw material or a change in design approach. The number of transformations available for a product characteristic may be finite or unlimited. We can construct trajectories that describe the potential states of a product evolution as we apply successive transformations with respect to technology forecasts. Each transformation may be repeated as many times as needed. Given an initial state S_0, the sequence of state vectors is represented by the following:

$$S_n = T_n(S_{n-1}).$$

The state-by-state transformations are then represented as $S_1 = T_1(S_0)$; $S_2 = T_2(S_1)$; $S_3 = T_3(S_2)$; …; $S_n = T_n(S_{n-1})$. The final state, S_n, depends on the initial state S and the effects of the actions applied at the intermediate stages.

STATE PERFORMANCE MEASUREMENT

The ability to measure performance is essential for the success of DEJI Systems Model. A measure of design performance can be obtained at each state of the transformation trajectories. The model can use a gain function associated with the kth design transformation. The gain specifies the magnitude of enhancement (e.g., time, quality, cost savings, revenue, equipment utilization) to be achieved by applying a given action. The difference between a gain and a performance specification is used as a criterion for determining design control actions. The performance deviation is defined as:

$$\delta = g^k (S) - p$$

where p is the performance specification. Given the number of transformations available and the current state vector, we can formulate a design policy, P, to represent the rule for determining the next action to be taken. In practice, managers do this subjectively, but effectively, even if not fully optimized. The better the experience and skills of the manager, the better the heuristic efficacy of the decision. In the mathematical framework, the total design gain is denoted as:

$$g(S \mid n, P) = g_1(S) + g_2(S) + \ldots \ldots g_n(S_{n-1}),$$

where n is the number of transformations applied and $g_i(.)$ is ith gain in the sequence of transformations. We can consider a design environment where the starting state vector and the possible actions (transformations) are specified. We have to decide what transformations to use in what order so as to maximize the total design gain. That is, we must develop the best design process. If we let $v(S \mid n)$ represent a quantitative measure of the value added by a design process based on the gain function described above, then the maximum gain will be given by the following expression:

$$v(S \mid n) = \max \{ r(S \mid n, P) \}$$

The maximization of the gain depends on all possible design rules that can be obtained with n transformations. Identification of the best design process will encompass both qualitative and quantitative measures in the design environment.

NON-DETERMINISTIC PERFORMANCE

In many design situations, the results of applying transformations to the design process may not be deterministic. The new state vector, the gain generated, or both may be described by random variables in accordance with semi-Markov processes. In such cases, we can define an expected total gain as the sum of the individual expected gains from all possible states starting from the initial state. If we

let S_p be a possible new state vector generated by the probabilistic process, and let
the following expression

$$P\left(S_p \mid S_i, T^k\right)$$

be the probability that the new state vector will be S_p, if the initial state vector is
S_i and the transformation T^k is applied. We can then write a recursive relation for
the expected total gain. That resulting complex mathematical expression is left to
the preference and desire of the reader of this section. In the application of this
approach, fuzzy modeling can also be incorporated for the heuristic evaluation of
the gain function without resorting to esoteric mathematical analysis. Practical
considerations related to the design phase of the DEJI Systems Model include the
following:

- Selection and validation of state variables
- Development of the method for generating the state equations (e.g., use of
  flow diagrams)
- Formulation of fuzzy model representation for probabilistic states
- Delineation of data input requirements for the state-space model
- Design of heuristic software tool (expert system) to govern the implementa-
  tion process

QUANTIFYING DESIGN PERFORMANCE

Based on the NSF-funded research project mentioned in Chapter 2, Sieger and
Badiru (1995) present a design performance quantification methodology.

The processes involved in the design of products have been the focus of interest for quite some time. While the general consensus suggests that such processes
lack structure and are not amenable to complete automation, it is also recognized
that much more can be done to assist designers in the creation of products. In
fact, it is reported that there are two situations where design automation tools can
be of benefit. The design process really consists of two distinct elements: logical
and creative. Although it is recognized that the capabilities of computers fall
short of that required for creative automation, there is considerable room for
improvement in the logical element. It is this element which the area of design
methodologies has been addressing since the early 1980s. These methods attempt
to reduce the demands placed on the designer by automating certain tasks, allowing designers to focus their efforts on creative aspects of the design process.
A problem which has received substantially less attention is one which considers
the incremental effect of decisions on the product being designed. Specifically,
this case considers how each subsequent decision combines with all previous
decisions to form an aggregate set, whose collective identity dictates the product's status in the progressive process toward the desired end product, service, or
result. To address this problem, a method for integrating the effects of design

tools, techniques, standards, participants, structure, and most importantly, customer desires, is required. Beginning with a design transformation approach, the performance parameter model is developed here in the context of the DEJI Systems Model.

As elucidated earlier in this chapter, the processes involved in the design of a product can be represented by a series of transformations. In this representation, each transformation takes an input and produces an output. The intent of each transformation selection is to advance the product toward a desired realization. The first step toward this realization is the knowledge of what transformations are appropriate. In addition to this transformation knowledge, the designer must also be aware of what initial conditions are imposed on the process. Hence, given a set of initial conditions and a list of appropriate transformations, the design process would proceed by sequentially applying selected transformations until some desired final state is reached or it is no longer advantageous to continue. In this approach, one can pursue a depth-first search or a breadth-first search.

To be able to intelligently select each design transformation, the design team must somehow be aware of the status of the product. This implies that the output of each design transformation provides useful information. If this were not the case, then it would be impossible to consider breaking the design process up into distinct segments. The best that could be done in that case would be to represent all product designs, no matter how complex, as a black box. Since it is possible to decompose the design process and recognize that the results of the latter stages build on those of the former stages (or else the process is random and or former stages are meaningless), it is logical to expect that the output of each design transformation provides useful information. Therefore, this information is valuable in capturing the essence of the product's status, along the spectrum of product development.

Visualizing the output of a transformation as a point in a decision space, it is clear that the series of decisions made during the design process would be represented by a series of points in that space with the composite of each point being a reflection of variables which are vital to the success of the design. Denoting the ith such variable as s_i, the design state vector corresponding to this n-dimensional construct can be formed by choosing n state variables from a set of measures:

$$S = \left(s_1,\ s_2,\ \ldots\ldots,s_n\right) \in s \subset M, \quad \forall s_i,\ i = 1,2,\ldots\ldots,\ n$$

where $M = \{\text{measures}\}$.

In this approach, M is considered to be the basic design variable set representing measures which are useful in capturing the essence of the product's status. Having established S, design actions can be applied whose effect is to modify one or more of the state variables contained in the vector. The only requirement in the application of these design actions is that they be selected from a valid set of transformations. For example, let the jth transformation applied in the design process be represented by T_j, and the finite or countably infinite set of valid design actions be represented as **H**:

$$T_j \in \mathbf{H}.$$

As indicated above, this example assumes that T_j is a member of **H**. Thus, application of this transformation to the state vector causes at least one of the s_i making up S to be modified. Given that the only restriction for application of T_j is that it be valid, a series of similar actions can be sequentially applied. In this series, each transform provides a functional mapping, $f(\ )$, between states. Designating the current design state as S_c and the resulting design state as S_R, the mapping between any two design states is provided by the following expression:

$$S_R = f\left(x|S_c\right) + \varepsilon$$

Since it is generally impossible to know the exact mapping between two stages, this relation has an associated error, e. In complex processes, the source of this error can result from an approximation about variable interactions or the omission of higher-order terms. In the design process, this approximation is initially due to difficulties in understanding customer desires, followed by conceptual uncertainty, and the relation of both to the resulting design. To minimize the effect of such error, a fuzzy approach was implemented by Sieger and Badiru (1995). Variations of this approach have previously been used in selecting parameter settings as well as alternative selections.

After identifying variables which can form a representative state vector, it is desirable to group these terms into classes. One such ordering may include technological, psychological, time-ordered, contractual, ethical, and economic groupings. However, after considering the processes involved in the design of a product, it appears that a more useful classification scheme exists. In this approach, state variables are grouped into performance, planning, and structural categories.

State variables falling into the performance group are those which allow the designer to have some immediate estimate of how well a component of the overall process is functioning. Planning measures indirectly determine what can be designed and developed, and have time, resource, and goal implications. The remaining category contains structural measures. These variables are essential in tracking the critical dimensions of components, which comprise design interfaces, including any customer-related structural specifications. Comparing these state variable groups with the Taguchi model, it is obvious that planning measures are analogous to control factors, performance measures are representative of indicative factors, structural factors correspond to signal factors, and customer uncertainty and process variability are characteristic of noise factors. Taguchi methods are statistical methods (i.e., robust design methods) developed by Genichi Taguchi to improve the quality of manufactured goods. The methods have also been applied generally to engineering, biotechnology, marketing, and advertising, in a way analogous to how the DEJI Systems Model, originally developed for product development, has been expanded to more diverse applications. Professional statisticians have welcomed the goals and improvements brought about by Taguchi methods, particularly by Taguchi's development of designs for studying variation.

The relationship between these three classes of state variables can be represented as interfacing block diagrams. Linking arrows between the planning and structural measures indicate that the planning measures determine what can be structurally designed. For example, given very little time and resources, it is expected that a structural design would run a greater risk of being both generic and poor, than it would, if given a vast amount of time and resources. In a similar manner, the structural measures determine the performance of the design. Lastly, the performance measures can be used to determine whether to update the planning measures.

STATE VARIABLE STRUCTURE

Consider the simple design scenario illustrated earlier in Figure 3.2. In this case, selection of appropriate design transformations proceeds by using a directed breadth-first heuristic. Following this heuristic at each of the intermediate design layers, various design transformations are considered until either a transformation, which is clearly desirable, is selected or the best among a known set is determined. To arrive at a desirable final state, the design team should select an efficient, ordered set of design actions. In this case, efficient selection implies that the design team is aware of the implications and extent of design actions, and can select them in an intelligent fashion. Having such knowledge permits avoidance of unnecessary reiteration of intermediate design stages. Since intelligent transformation selection is largely dictated by an ability to integrate information and resources, an automated tool for integrating such factors is desirable. As a part of integration, this tool should also determine an associated performance rating. This rating could then be used during the concurrent engineering process as a mechanism for guidance. Using the Triple C model as a template for integration efforts, this automated tool could then serve as a vehicle for the reduction of no efficient use of both information and resources. Thus, by aggregating key design factors, this tool could be used by the design team to avoid counterproductive activities. Generalizing the example presented in Figure 3.2, the current performance parameter model accounts for three common design process stages (i.e., specification and planning, conceptual, and product design stages).

The relationship between these three stages and the key state variables is depicted in Figure 3.3. In this figure, the state variables are shown in bold italics. The current model adds a fourth stage. The purpose of this stage is to provide direct comparisons between the current product and benchmarks. Also, instead of referring to each phase as a stage, they are referred to as flows. Elements which

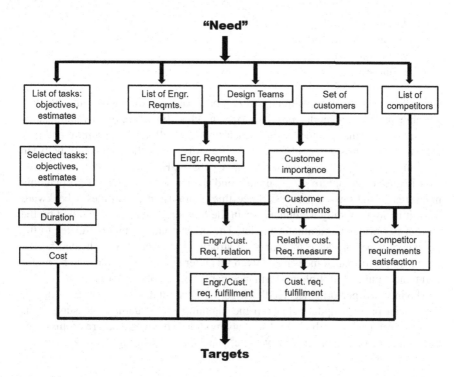

FIGURE 3.3   Specification and planning flow.

directly precede any node are said to "drive" the node. The purpose for this distinction is to emphasize the hierarchical nature which exists between the state variables comprising each design stage, as well as those between stages.

Classifying the state variables into the performance, planning, and structural measure groupings results in the following summary.

1. Performance measures:
   a. engineering/customer requirement rela-tion
   b. relative customer requirement measure
   c. engineering/customer requirement fulfill-ment
   d. customer requirement fulfillment
   e. competitor requirement satisfaction
   f. functional satisfaction
   g. customer satisfaction
   h. concept desirability
   i. combination desirability
2. Planning measures:
   a. objectives
   b. estimates
   c. duration

d. cost
e. customer importance
3. Structural measures:
a. targets

Using this classification summary to compare the flow diagrams with the state variable interaction model provides two insights. First, although relationships between the planning and structural measures, as well as the structural and performance measures, are contained in the flow diagrams, the connection between performance and planning measures appears absent. However, this apparent omission is due to the nature of the update and the early occurrence of all planning measures in the design process. Since this relationship constitutes a backward flow, it is not explicitly accounted for in the flow diagrams but is handled by the current performance parameter model. The other point of interest relates to the influence of performance measures on structural measures, as indicated by the cluster of performance measures. Given this interaction, the system can be modified to account for multiple performance parameter feedback forms.

Having two performance feedback forms necessitates inclusion of a device which can provide updates to both the planning and structural measures. This device is referred to as the hybrid strategy module. The hybrid nature of this module results from combining fuzzy logic, state-space modeling, and performance metric techniques.

PERFORMANCE QUANTIFICATION AND METRIC

During product development, difficulties occur when there is uncertainty about the applicability and effects, at both global and local levels, of design actions. Although it is important to make informed decisions at each stage in this process, this need is heightened early in the design process, given the significant financial commitment and ambiguity involved. To assist the design team in making quality decisions during these and all other stages comprising the design process, a performance metric was developed. This metric serves to calculate a performance rating based on the identified state vector, S. Using the state variables provided in the previous section as a minimum set, the form of the overall performance metric is as follows:

$$P_T(t) = \frac{1}{w_N} \sum_{i=1}^{N} w_{s_i} P_{s_i}(t),$$

where

$$w_N = \sum_i w_{s_i}.$$

As indicated in the above model, each of the i design process stages has an associated performance rating, P_{si} and weight, w_{si}. Hence, at any point during product development, the overall performance rating equates to the weighted sum of the N individual stages. Corresponding to each of the individual design stages is the following stage performance metric:

$$P_{S_i}(t) = e^{a_{si}t} \sum_{j=1}^{M_{S_i}} A_j$$

In this metric, the A_j represents design actions which have been selected from a set of valid ith stage transformations. The a_{si} represents a stage weight and is inversely related to the w_{si} contained in the overall performance metric. As an example of the utilization of these stage weights, consider their use in the current performance parameter model. In this model, when the elapsed design time, t, resides within a specified stage range, $t_{lower} \leq t < t_{upper}$, a_{si} approaches 0. Otherwise, a_{si} approaches $-\infty$. The result of this weighted assignment is for the exponential multiplier to implement a pulsed operation. Hence, this approach sets up a performance window which limits the effects of design actions to the current stage. The effect of each invoked stage action on the product being designed is then taken into account using the following relation:

$$A_j = \frac{1}{6}(\gamma + 3\upsilon)$$

where

$$L = \frac{1}{m}\sum_{i=1}^{m} e^{a_{nom}(q_i - d_i)^2} + \frac{1}{n}\sum_{j=1}^{n} e^{a_{lg}(1 - d_j)^2} + \frac{1}{p}\sum_{k=1}^{p} e^{a_{sm}d_k^2}$$

$$\upsilon = \frac{1}{q^{N(e-1)}}\sum_{i=1}^{q}\sum_{k=1}^{N}\left(e^{\mu_i(n_j)} - 1\right)$$

In this formulation, each of the selected actions can impact loss, L, and vagueness, V, terms. These two terms have been utilized since they are associated with the two dominant forms of uncertainty, the former being dominant late in the design process, while the latter is dominant early. The three types of loss that are considered are referred to as the nominal the better (a_{nom}), the larger the better, (a_{lg}), and the smaller the better (a_{sm}).

HYBRID PERFORMANCE STRATEGY

The hybrid performance strategy consists of the following:

- Planning measures
- Performance/Quality measures
- Structural measures

A subset of the design state variables (i.e., the performance measures) provides two types of information. One type is useful in providing recommendations for revisions to planning measures and the other toward structural measures. To handle transmission of this information to the design team, a hybrid strategy module was developed by Sieger and Badiru (1995). The role of this module is to interpret the status of the design and make recommendations to the design team. To assist in providing these recommendations, a second-order servo controller is implemented as a front-end processor for performance aggregation. Using the performance quantification metric defined previously, individual design variant inputs have the following general time-domain representation:

$$u(t) = de^{st}u(t-b)$$

where a and d are stage constants, and b is a time shift factor. This generic step input is then transferred, along with the transfer function, into the s domain to obtain the frequency response of the controller to a single input. Bringing the controller response back to the time domain and performing necessary operations results in the following critically damped time-domain controller response:

$$y_i(t) = 0, \quad t < b,$$

$$y_i(t) = \begin{cases} \dfrac{d_1 w_n^2}{(a+w_n)^2} e^{a_l(t-b_j)} - e^{-w_n(t-b_j)} \left[\dfrac{d_i w_n^2 (t-b_j)}{w_n - a} + \dfrac{d_i w_n^2}{(-w_n - a)^2} \right], & t \geq b, \end{cases}$$

where w_n is the undamped natural frequency. This critically damped form has been chosen to minimize overshoot and oscillation of the output. To use the controller in the design environment, the single-input-single-output (SISO) model is extended to handle multiple inputs. In this case, each of the i design actions has the corresponding indexed form:

$$u_i(t) = \begin{cases} 0 & t < b_i \\ d_i e^{a_i t} u(t-b) & t \geq b_i \end{cases}$$

As in the SISO model, design variants are treated as step inputs. The corresponding output response of the controller is obtained using the principle of superposition. The combined effect of these actions, as shown below, is designated as y_T:

$$y_T = \sum_{i=1}^{N} y_i(t) \Rightarrow \text{actual performance} = p_a$$

Since accumulation of the effects of design actions conveys the status of the state vector, and this state vector was formed by selecting variables which capture the essence of the product's status, y_T equates to the actual performance, P_a, of the design. Comparing the actual performance of the design to a specified desired level, a performance deviation is obtained:

$$\delta = P_a - P_d.$$

This value of δ is then used in conjunction with n, the number of transforms which have been applied, and t, the elapsed design time, to make adjustments to a fuzzy associative memory (FAM). The FAM resides within the hybrid strategy module and is used to provide recommendations regarding subsequent activation of actions. Adjustments to the FAM are done in accordance with an operational design policy, P, so that the recommendations provided by this module maximize the value added to the design. As an example of P, consider the marketing slogan that says Ford trucks are "Built Ford tough." In this case, FAM rule adjustments would be made so that the toughness measures of the product would be emphasized. Given such a policy, calculation of the value-added measure, $v(\ )$, is obtained by using the following equation:

$$v\left(S \mid n,t\right) = \max_p \left\{r\left(S \mid n,\delta,t\right)\right\}$$

In this equation, $r(\ )$ represents FAM rule activation. Once a particular action is selected, the gain associated with this action is determined using the following relation:

$$g_i\left(S_{i-t}\right) = g_t\left(t\right) + g_n\left(A\right)$$

The first component of this relation accounts for the time spent performing the activity relative to the time which has been allocated to this stage. This ratio is then weighted by the performance rating which was achieved at the end of the activity:

$$g_1\left(t\right) = \frac{t_{s_i} - t_{s_{i-1}}}{T_{s_i}} p_i$$

The second gain component focuses on activity completion and has the following form:

$$g_2\left(A\right) = \begin{cases} w_{\text{act}} & \text{activity completed} \\ 0 & \text{activity not completed} \end{cases}$$

Provided that the activity has been completed, a value between zero and one is assigned to w_{act}. This weight serves as a measure for completeness of the activity, where a value of 1 means that nothing remains to be done on that activity. The result of each selected action is then added to all previous actions to obtain an overall gain associated with the product. As indicated below, this summation differs from the way the overall performance measure is determined, given its dependence on the n previous actions and current performance deviation:

$$g\left(s \mid n, \delta\right) = \sum_{i-t}^{n} g_t\left(S_{i-t}\right)$$

Summarizing the flow of information into and out of the hybrid strategy, a flow chart is used to illustrate the use of this module within the context of the performance parameter model and design team preferences. The design team is provided with four types of information. The three types which result from operation of the hybrid module have been discussed above. The fourth type is provided by the servo controller. Given these inputs, the design team makes its decision regarding which action should be selected. Selection of the action provides the input generated to the controller. This cycle continues throughout the design process.

As an illustration of the use of the performance parameter model, consider the following hypothetical design scenario. For this example, the set of valid design actions is given in Table 3.1. In this table, each transformation listing consists of its stage classification (i.e., S = specification and planning stage, C = conceptual design stage, and P = product design stage), unique identification, and brief description.

USING PUGH METHOD FOR DESIGN CONCEPT SELECTION

The Pugh method is a decision-matrix approach for design concept selection. Invented by Stuart Pugh (Pugh, 1991), the method is a qualitative technique used to rank the multi-dimensional options of an option set. It is frequently used in engineering for making design decisions but can also be used to rank investment options, vendor options, product options, or any other set of multi-dimensional alternatives. It is also called by a variety of names, including Pugh analysis, decision-matrix method, decision matrix, decision grid, selection grid, selection matrix, problem matrix, problem selection matrix, problem selection grid, solution matrix, criteria rating form, criteria-based matrix, and opportunity analysis. Literature references to the Pugh Method can be found in Lønmo and Muller (2014), Muller et al. (2011), Onn et al., (2020), Raudberget (2010), Villanueva et al., (2016), and Wurthmann (2020). The Pugh Method (Pugh et al., 1996) can be used in parallel with the analytic hierarchy process (AHP) for pair-wise comparison of design alternatives (McCauley-Bell and Badiru, 1996).

To start the design process, a "need" must be identified. To establish such a need, Action 1, *define overall objective,* is performed. Given this need, a design team is formed. At this point, an initial list of objectives is established. This list includes both short- and long-term tasks which must be performed. The short-term objectives include identification of need, problem statement, design cost, product cost, concept design time, completion date, competitors, customer(s),

TABLE 3.1

List of Valid Design Actions

Design ID	Stage Code	Description	Design ID	Stage Code	Description
1	S	Define overall objective	2	S	Refine overall objective
3	S	Produce problem statement	4	S	Refine problem statement
5	S	Develop task list	6	S	Refine task list
7	S	Solicit customer requirements	8	S	Refine customer requirements
9	S	Assign requirement weights	10	S	Revise requirement weights
11	S	Assess customer requirement expectations	12	S	Select product competitors
13	S	Select engineering requirements	14	S	Refine engineering requirements
15	S	Perform benchmarking	16	S	Weight engineering/customer requirements
17	S	Determine engineering targets	18	C	Determine overall functional objective
19	C	Decompose function objective into primitive	20	C	Generate concepts
21	C	Perform preliminary concept evaluation	22	C	Perform modified Pugh Method
23	C	Determine set desirability	24	C	Determine combination desirability
25	P	Vary effects	26	P	Vary effectors
27	P	Form initial layout	28	P	Dimension part(s)
29	P	Analyze layout	30	P	Revise initial drawing
31	P	Produce detailed drawing	32	P	Produce bill of materials

customer requirements, engineering requirements, engineering/customer requirement relation, relative customer requirement measure, competitor requirement satisfaction, and targets. All of these items represent tasks that must be completed during the specification and planning stage. Long-term objectives correspond to tasks which are to be addressed in subsequent design stages.

Objectives contained on the initial task list are treated by the performance parameter model as larger-the-better loss terms. In addition to having this type of loss, many of the terms also have a vagueness component. This component occurs whenever the design team uses a fuzzy membership function as a way to represent variable uncertainty. Since the fuzzy representation captures the designer's perception of what the variable values should be, and these perceptions evolve over the course of the design process, the fuzzy forms are similarly adapted. This evolutionary process is depicted in typical fuzzy model diagrams (triangles, rectangles, trapezoids, etc.). In such diagrams, T relates to the numeric variable level which the design team perceives as desirable (i.e., $\mu_T = 1.0$). Given that the need has been established and it is on the task list, the resulting performance rating can be determined using the following reduced metric:

$$P_{s_1} = \frac{1}{14}\sum_{j-1}^{14} e^{a_{1g}(t-d_j)^2}$$

Since all but 1 of the 14 tasks have not yet been performed, and need does not have a vagueness component, $P_T = 0.012$. Continuing with product development, Action 3, *produce problem statement,* is invoked, resulting in a new performance level, $P_T = 0.024$. Assessing the effect of the next action, Action 5, is somewhat more involved, because this transformation generates variables, which have fuzzy representations. Specifically, estimates of design cost, product cost, concept design time, completion date, and later, customer requirements (Action 7), engineering/customer requirement relation (Action 16), relative customer measure (Action 11), competitor requirement satisfaction (Action 15), and targets (Action 17), all use a fuzzy representation. To simplify this example, only one of these factors, *concept design time,* will be used in the remainder of this illustrative example. Working with only concept design time and the loss terms, the overall performance rating will be carried through the end of the conceptual design stage (i.e., when the fuzzy parameter is fixed). Since these few terms do not account for all of those that would normally be considered, the overall performance ratings will be referred to using a prime notation. Thus, at the end of Action 3, $P_T = 0.024$.

Applying Action 5, it is determined that the initial estimate for concept design time is captured by a triangular fuzzy number (TFN). See Sieger and Badiru (1995) for the figures representing the fuzzy set plots. Using set notation, this TFN is represented as (30, 40, 40). In this case, the interpretation is that the designer anticipates that the conceptual design must be completed by the 40th day. Using set notation, this TFN is represented as (30, 40, 40). In this case, the interpretation is that the designer anticipates that the conceptual design must be completed by the 40th day. In addition, the design team suspects that there will be some incentive for finishing up to 10 days before the deadline. Setting $N = 20$, the vagueness component is determined, $V = 0.393$. Consequently, $P_T = 0.268$.

During Action 7, the customer indicates that the incentive will only be applicable to 5 days prior to the deadline. Having this new information, Action 6 is applied. This activity results in a new concept design-time representation of (32, 38, 40). Thus, after taking the incentives into account, it was determined that the resulting conceptual design should ideally be delivered on the 38th day. Given this form, the corresponding performance metric is $P_T = 0.305$.

An overall performance metric summary of actions taken up to this point, as well as those which followed, is plotted graphically. For simplicity in determining the controller response, the time between actions and the undamped natural frequency were set as follows:

$$b_{i+1} = b_i + 1; \quad b_0 = 0; \quad w_n = 1$$

Furthermore, letting $T_{s1}-T_{s2}$ results in the associated action gains. Unity weights were assigned to the completion of each activity. Much of the nonlinearity that would have otherwise been present has been eliminated due to the simplifying assumptions which have been made. Finally, as an illustration of the effect of the value-added and policy parameters, consider the decision which was made by the design team when they received the incentive information from the customer. Had the design policy dictated that all available time be used instead of being influenced by the incentives, the TFN corresponding to concept design time might have been (35, 40, 40). Selection of such a form would have been reflected in the recommendation provided by the hybrid strategy module. The point of going through the comprehensive quantitative illustration in this example is to demonstrate that design processes can be quantified and analytically assessed. In practice, rarely would a designer resort to this level of esoteric manipulation. Thus, a simplified mix of heuristics, qualitative, and simple mathematical representations can be used. The point of the DEJI Systems Model is to provide a rigorous framework over which design decisions are made for the purpose of facilitating the ease of the stages that follow, such as system evaluation, system justification, and system integration.

The design performance parameter model presented in this chapter represents an effective tool for evaluating the status of a product during development. Using a servo controller, the effects of actions taken during the design process are aggregated into a performance rating. In addition to enabling aggregation, this approach facilitates operator tuning, provides a consistent modular structure supporting the natural expression of the design states, and continuously outputs the current status of the product. Utilizing a hybrid strategy module, the performance parameter model also provides gain and value-added measures, which can be used for process control. Combining these features, the performance parameter model is able to provide valuable guidance information to the design team in any organization endeavor. Even if mathematical representations are not used, designers and analysts still need to be cognizant of the considerations and nuances that exist in the system environment. This is the framework or platform that DEJI Systems Model provides.

REFERENCES

Badiru, A. B. (2008), *Triple C Model of Project Management: Communication, Cooperation, and Coordination*, Taylor & Francis Group/CRC Press, Boca Raton, FL.

Badiru, A. B. (2012), "Application of the DEJI Model for Aerospace Product Integration," *Journal of Aviation and Aerospace Perspectives (JAAP)*, Vol. 2, No. 2, pp. 20–34, Fall 2012.

Badiru, Ibrahim A., and M. W. Neal (2013), "Use of DFSS Principles to Develop an Objective Method to Assess Transient Vehicle Dynamics," *SAE International*, Vol. 25, No. 4, pp. 1–8.

Lønmo, L., and G. Muller (2014). "Concept Selection: Applying Pugh Matrices in the Subsea Processing Domain," In *Proceedings of INCOSE International Symposium*, Rome, Italy, Vol. 24, No. 1, pp. 583–598.

McCauley-Bell, Pam, and A. B. Badiru, "Fuzzy Modeling and Analytic Hierarchy Processing to Quantify Risk Levels Associated with Occupational Injuries – Part I: The Development of Fuzzy Linguistic Risk Levels," *IEEE Transactions on Fuzzy Systems*, Vol. 4, No. 2, May 1996, pp. 124–131.

Muller, G., D. G. Klever, H. H. Bjørnsen, and M. Pennotti (2011), "Researching the Application of Pugh Matrix in the Sub-Sea Equipment Industry," In *2011 INCOSE Conference on Systems Engineering Research*, Redondo Beach, CA, pp. 8–14.

Onn, Choo Wou et al (2020), "Selection of Pipeline Investigation Robot via Pugh Method," *Proceedings of International Conference on Innovation and Technopreneurship, 2020 INTI Journal*, eISSN:2600-7320, Vol. 2020, No. 025, pp. 9–16.

Peacock, Brian (2009), *The Laws and Rules of Ergonomics in Design*, EAZI Printing, Amelia Island, FL.

Peacock, Brian (2019), *Human Systems Integration*, EAZI Printing, Amelia Island, FL.

Pugh, S. (1991), *Total Design: Integrated Methods for Successful Product Engineering*. Addison-Wesley, New York, NY. ISBN 0-201-41639-5

Pugh Stuart, Don Clausing, and Ron Andrade, (1996), *Creating Innovative Products Using Total Design*, Addison-Wesley, New York, NY. ISBN 0-201-63485-6

Raudberget, D. (2010), "The Decision Process in Set-Based Concurrent Engineering: An Industrial Case Study," In *DS 60: Proceedings of DESIGN 2010, the 11th International Design Conference*, Dubrovnik, Croatia (pp. 937–946).

Sieger, David B., and A. B. Badiru (1995), "A Performance Parameter Model for Design Integration," *Engineering Design and Automation*, Vol. 1, No. 3, pp. 137–148.

Villanueva, P. M., R. Lostado Lorza, and M. Corral Bobadilla (2016), "Pugh's Total Design: The Design of an Electromagnetic Servo Brake with ABS Function: A Case Study," *Concurrent Engineering*, 24(3), 227–239.

Wurthmann, K. (2020), "Conducting Pugh Method-based Trade Studies during Product Development: The case of evaluating Turbofan versus Turboprop versus Piston Engine Alternatives for UAVs," In *AIAA Scitech 2020 Forum*, Atlanta, GA, pp. 705–711.

4 Evaluation in DEJI Systems Model®

INTRODUCTION

Evaluation, in the context of DEJI Systems Model®, goes beyond the typical feasibility analysis. It is at the evaluation stage that design trade-offs can be assessed, compared, and, possibly, optimized. In carrying out evaluation in DEJI Systems Model, we will use a mixture of both human and digital for assessing the system of interest. Although many of the recommended techniques of evaluation are drawn from familiar quantitative repertoire, many times, the evaluation becomes a matter of subjective perception. To simplify the concept of evaluation for the purpose of DEJI Systems Model, this chapter suggests selected metrics and options that can be adapted for the evaluation stage of DEJI Systems Model. Once again, we iterate the sequence of the stages of the model as design, evaluation, justification, and integration. Anything can be designed, but can it pass the test of evaluation? Can it be justified for the organization? Can it be integrated into the expectations and realities of the operating environment with regard to people, process, tools availability, and timing. Due to the 2020 emergence of COVID-19 pandemic, the world has seen many instances of disconnects between plans, actions, timing, and acceptability. This is evident in the challenges that developed with testing, vaccination, and treatment campaigns and mandates. As long as the pandemic remains with us, the struggle will continue, unless we adopt a more systems-based approach to resolving the worldwide problem. In this regard, the DEJI Systems Model is advocated for imparting a discernible structure to what needs to be done. Figure 4.1 presents the stage-by-stage overlap structure of DEJI Systems Model. Each user can determine the specific methodologies to be employed for each stage of the model. This permits a flexible mix of quantitative and qualitative (hard and soft) tools and techniques in using the model. Badiru and Sieger (1998), Badiru and Omitaomu (2007), and references therein present several quantitative and qualitative methods suitable for consideration by users of DEJI Systems Model.

COST AND VALUE OF A SYSTEM

Economic analysis is just one of the primary methods for assessing the worthiness of a system under consideration. Each organization may have its own internal value stream process. The point of DEJI Systems Model is to elicit a serious and rigorous process of evaluating whatever is being contemplated. Industrial

Stages may overlap depending on organizational needs, preferences, and organic tools and techniques.

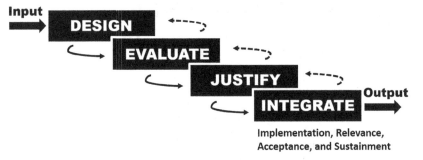

FIGURE 4.1　Stage-by-stage overlap structure of DEJI Systems Model.

enterprises have fundamentally unique characteristics and require unique techniques of economic analysis. Thus, while the methodologies themselves may be standard, the specific factors or considerations may be industrially focused. Fortunately, most of the definitions used in general economic analyses are applicable to system or industrial economic analysis. This chapter presents computational definitions, techniques, and procedures for economic analysis. As in many cases, a project basis is used for the evaluation of systems. So, for the purpose of the presentations in this chapter, "project" and "system" may be used interchangeably. To put things within a focused context, an industrial enterprise is used as the illustrative basis for the discussions in the sections that follow.

We need to define and clarify some basic terms often encountered in economic analysis. Some terms appear to be the same but are operationally different. For example, the term "economics" must be distinguished from the term "economic analysis," and even more specifically from the term "engineering economic analysis." "Economics" is the study of the allocation of the scarce assets of production for the purpose of satisfying some of the needs of a society. "Economic analysis," in contrast, is an integrated analysis of the qualitative and quantitative factors that influence decisions related to economics. Finally, an "engineering economic analysis" is an analysis that focuses on the engineering aspects. Examples of the engineering aspects typically considered in an economic design process include the following:

- Product conceptualization
- Research and development
- Design and implementation
- Prototyping and testing
- Production
- Transportation and delivery

Industrial economics is the study of the relationships between industries and markets with respect to prevailing market conditions, firm behavior, and economic performance. In a broader sense, the discipline of industrial economics focuses on a broad mix of industrial operations involving real-world competition, market scenarios, product conceptualization, process development, design, pricing, advertising, supply chain, delivery, investment strategies, and so on. While this chapter focuses on some of the cost aspects, the full range of industrial economics is beyond the scope of the chapter. Readers can adopt and adapt the diverse computational techniques that are available in performing economic analysis in industrial, project, or system settings.

A design or system can be evaluated on the basis of earned value analysis. "Earned value analysis" is often used in industrial project economic analysis to convey the economic status of a project. "Planned value" (PV) refers to the portion of the approved cost that is planned to be spent during a specific period of the project. "Actual cost" (AC) is the total direct and indirect costs incurred in accomplishing work over a specific period of time. "Earned value" is defined as the budget for the work accomplished in a given period. Formulas relating to these measures are used to assess the overall economic performance of a project. Specific definitions are presented below:

- *Cost variance* (CV) equals *earned value* (EV) minus *actual cost* (AC). The cost variance at the end of the project is the difference between the budget at completion (BAC) and the actual amount spent:

$$CV = EV - AC >$$

  - A positive CV value indicates that costs are below budget.
  - A negative CV value indicates a cost overrun.

- *Schedule variance* (SV) equals earned value (EV) minus *planned value* (PV). Schedule variance will ultimately equal zero when the project is completed because all of the planned values will have been earned:

$$SV = EV - PV.$$

  - A positive SV value indicates that a project is ahead of schedule.
  - A negative SV value indicates that the project is behind schedule.

- *Cost performance index* (CPI) equals the ratio of EV to AC. A CPI value less than 1.0 indicates a cost overrun of estimates. A CPI value greater than 1.0 indicates a cost under-run of estimates. CPI is the most commonly used cost-efficiency indicator:

$$CPI = EV / AC.$$

  - A CPI greater than 1.0 indicates costs are below budget.
  - A CPI less than 1.0 indicates costs are over budget.

- *Cumulative CPI* (CPI^C) is used to forecast project costs at completion. CPI^C equals the sum of the periodic earned values (EV^C) divided by the sum of the individual actual costs (AC^C):

$$CPI^C = EV/AC^C.$$

- A *schedule performance index* (SPI) is used, in addition to the schedule status, to predict completion date and is sometimes used in conjunction with CPI to generate project completion estimates. SPI equals the ratio of the EV to the PV:

$$SPI = EV/PV.$$

- An SPI greater than 1.0 indicates a project is ahead of schedule.
- An SPI less than 1.0 indicates a project is behind schedule.

- *Budget at completion* (BAC), *actual cost* (AC^C) to date, and *cumulative cost performance* index (CPIC) are used to calculate the *estimated total cost* (ETC) and the *estimated actual cost* (EAC), where the BAC is equal to the total planned value (PV) at completion for a scheduled activity, work package, control account, or other WBS component:

$$BAC = total\ cumulative\ PV\ at\ completion.$$

- The *estimated total cost (ETC)*, based on atypical variances, is an approach that is often used when current variances are seen as atypical, and the project management team expects that similar variances will not occur in the future. ETC equals the BAC minus the cumulative earned value to date (EV^C):

$$ETC = \left(BAC - EV^C \right).$$

- *Estimate at completion* (EAC) is the expected total project cost upon completion with respect to the present time. There are alternate formulas for computing EAC depending on different scenarios. In one option, EAC equals the actual costs to date (AC^C) plus a new ETC that is provided by the project organization. This approach is most often used when past performance shows that the original estimating assumptions are no longer applicable due to a change in conditions:

$$EAC = AC^C + ETC.$$

- *EAC using remaining budget.* EAC equals AC^C plus the budget required to complete the remaining work, which is the budget at completion (BAC) minus the earned value (EV). This approach is most often used when current variances are seen as atypical and the project management team expects similar variances will not occur in the future.

$$EAC = AC^C + BAC - EV.$$

- *EAC using CPI^C*. EAC equals actual costs to date (AC^C) plus the budget required to complete the remaining project work, which is the BAC minus the EV, modified by a performance factor (often the CPI^C). This approach is most often used when current variances are seen as typical of future variances:

$$EAC = AC^C + \left((BAC - EV)/CPI^C\right).$$

- *Present Value* (PV) is the current value of a given future cash-flow stream, discounted at a given rate. The formula of how to calculate a present value follows:

$$PV = FV/(1+r)^{(n)}.$$

EVALUATION FOR RESOURCE ALLOCATION

Worker assignment is one of the basic responsibilities of a system. A well-designed system that cannot be sustained with an efficient and effective assignment of workers may not serve the interest of the organization fully. Such a system will be evaluated poorly within the application of DEJI Systems Model. Operations-research techniques are often used to enhance resource allocation decisions in engineering and industrial projects. One common resource-allocation methodology is the resource-assignment algorithm. This algorithm can be used to enhance the quality of resource-allocation decisions. Suppose there are n tasks which must be performed by n workers. The cost of worker i performing task j is c_{ij}. It is desired to assign workers to the tasks in a fashion that minimizes the cost of completing the tasks. This problem scenario is referred to as the assignment problem. The technique for finding the optimal solution to the problem is called the assignment method. The assignment method is an iterative procedure that arrives at the optimal solution by improving on a trial solution at each stage of the procedure.

The assignment method can be used to achieve an optimal assignment of resources to specific tasks in an industrial project. Although the assignment method is cost-based, task duration can be incorporated into the modeling in terms of time-cost relationships. The objective is to minimize the total cost of the project. Thus, the formulation of the assignment problem is as shown below:

Let

$x_{ij} = 1$ if worker i is assigned to task j, $j = 1, 2, \ldots, n$,
$x_{ij} = 0$ if worker i is not assigned to task j

c_{ij} = cost of worker i performing task j.

$$\text{Minimize:} \qquad z = \sum_{i=1}^{n}\sum_{j=1}^{n} c_{ij} x_{ij}$$

$$\text{Subject to:} \qquad \sum_{j=1}^{n} x_{ij} = 1, \qquad i = 1, 2, \ldots, n$$

$$\sum_{i=1}^{n} x_{ij} = 1, \qquad j = 1, 2, \ldots, n$$

$$x_{ij} \geq 0, \; i, j = 1, 2, \ldots, n$$

The above formulation uses the non-negativity constraint, $x_{ij} \geq 0$, instead of the integer constraint, $x_{ij} = 0$ or 1. However, the solution of the model will still be integer-valued. Hence, the assignment problem is a special case of the common transportation problem in operations research, with the number of sources (m) = number of targets (n), $S_i = 1$ (supplies), and $D_i = 1$ (demands). The basic requirements of an assignment problem are these:

1. There must be two or more tasks to be completed;
2. There must be two or more resources that can be assigned to the tasks;
3. The cost of using any of the resources to perform any of the tasks must be known;
4. Each resource is to be assigned to one and only one task.

If the number of tasks to be performed is greater than the number of workers available, we will need to add *dummy workers* to balance the problem. Similarly, if the number of workers is greater than the number of tasks, we will need to add *dummy tasks* to balance the problem. If there is no problem of overlapping, a worker's time may be split into segments so that the worker can be assigned more than one task. In this case, each segment of the worker's time will be modeled as a separate resource in the assignment problem. Thus, the assignment problem can be extended to consider partial allocation of resource units to multiple tasks.

The assignment model is solved by a method known as the Hungarian method, which is a simple iterative technique. Details of the assignment problem and its solution techniques can be found in operations-research texts. As an example, suppose five workers are to be assigned to five tasks on the basis of the cost matrix presented in Table 4.1. Task 3 is a machine-controlled task with a fixed cost of $800 regardless of the specific worker to whom it is assigned. Using the assignment method, we obtain the optimal solution presented in Table 4.2, which indicates the following:

$$x_{15} = 1, x_{23} = 1, x_{31} = 1, x_{44} = 1, \text{ and } x_{52} = 1.$$

Thus, the minimum total cost is given by

$$\text{TC} = c_{15} + c_{23} + c_{31} + c_{44} + c_{52} = \$\left(400 + 800 + 300 + 400 + 350\right) = \$2,250.$$

TABLE 4.1
Cost Matrix for Resource Assignment Problem

Worker	Task 1	Task 2	Task 3	Task 4	Task 5
1	300	200	800	500	400
2	500	700	800	1250	700
3	300	900	800	1000	600
4	400	300	800	400	400
5	700	350	800	700	900

TABLE 4.2
Solution to Resource Assignment Problem

Worker	Task 1	Task 2	Task 3	Task 4	Task 5
1	0	0	0	0	1
2	0	0	1	0	0
3	1	0	0	0	0
4	0	0	0	1	0
5	0	1	0	0	0

SYSTEM-BASED EVALUATIVE ECONOMICS

In industrial operations that are subject to risk and uncertainty, probability information can be used to analyze resource utilization characteristics of the operations. Suppose the level of availability of a resource is probabilistic in nature. For simplicity, we will assume that the level of availability, X, is a continuous variable whose probability density function is defined by $f(x)$. This is true for many resource types, ranging from funds and natural resources to raw materials. If we are interested in the probability that resource availability will be within a certain range of x_1 and x_2, then the required probability can be computed as follows:

$$P(x_1 \leq X \leq x_2) = \int_{x_1}^{x_2} f(x) dx.$$

Similarly, a probability density function can be defined for the utilization level of a particular resource. If we denote the utilization level by U and its probability density function by $f(u)$, then we can calculate the probability that the utilization will exceed a certain level, u_0, by the following expression:

$$P(U \geq u_0) = \int_{u_0}^{\infty} f(u) du.$$

Suppose that a critical resource is leased for a large project. There is a graduated cost associated with using the resource at a certain percentage level U. The cost is specified as $10,000 per 10% increment in utilization level above 40%. A flat cost of $5,000 is charged for utilization levels below 40%. The utilization intervals and the associated costs are presented below:

$$U < 40\%, \$5,000$$
$$40\% \leq U < 50\%, \$10,000$$
$$50\% \leq U < 60\%, \$20,000$$
$$60\% \leq U < 70\%, \$30,000$$
$$70\% \leq U < 80\%, \$40,000$$
$$80\% \leq U < 90\%, \$50,000$$
$$90\% \leq U < 100\%, \$60,000.$$

Thus, a utilization level of 50% will cost $20,000, while a level of 49.5% will cost $10,000. Suppose the utilization level is a normally distributed random variable with a mean of 60% and a variance of 16% squared, and that we are interested in finding the expected cost of using this resource. The solution procedure involves finding the probability that the utilization level will fall within each of the specified ranges. The expected value formula will then be used to compute the expected cost as shown below:

$$E[C] = \sum_{k} x_k P(x_k)$$

where x_k represents the kth interval of utilization. The standard deviation of utilization is 4%. Thus, we have the following:

$$P(U < 40) = P\left(z \leq \frac{40-60}{4}\right) = P(z \leq -5) = 0.0$$
$$P(40 \leq U < 50) = 0.0062$$
$$P(50 \leq U < 60) = 0.4938$$
$$P(60 \leq U < 70) = 0.4938$$
$$P(70 \leq U < 80) = 0.0062$$
$$P(80 \leq U < 90) = 0.0$$
$$E(C) = \$5000(0.0) + \$10,000(0.0062) + \$20,000(0.4938)$$
$$+ \$30,000(0.4938) + \$40,000(0.0062) + \$50,000(0.0)$$
$$= \$25,000$$

Based on the above calculations, it can be expected that leasing this critical resource will cost $25,000 in the long run. A decision can be made as to whether to lease the resource, buy it, or substitute another resource for it based on the information gained from this calculation.

THE MARR APPROACH

Minimum Annual Revenue Requirement Analysis (MARR Analysis) (See Badiru and Russell, 1987) is another possible technique for evaluating alternatives (or design options). Companies evaluating capital expenditures for proposed projects must weigh the expected benefits against the initial and expected costs over the life cycle of the project. One method that is often used is the minimum annual revenue requirement (MARR) analysis. Using the information about costs, interest payments, recurring expenditures, and other project-related financial obligations, the minimum annual revenue required by a project can be evaluated. We can compute the break-even point of the project. The break-even point is then used to determine the level of revenue that must be produced by the project in order for it to be profitable. The analysis can be done with either the *flow-through* method or the *normalizing method.*

The factors to be included in MARR analysis are initial investment, book salvage value, tax salvage value, annual project costs, useful life for book-keeping purposes, book depreciation method, tax depreciation method, useful life for tax purposes, rate of return on equity, rate of return on debt, capital interest rate, debt ratio, and investment tax credit. This section presents an overall illustrative example of how companies use MARR as a part of their investment decisions. The minimum annual revenue requirement for any year n may be determined by means of the net cash flows expected for that year:

Net Cash Flow = Income − Taxes − Principal Amount Paid.

That is,

$$X_n = (G - C - I) - I - P,$$

where
  X_n = annual revenue for year n
  G = gross income for year n,
  C = expenses for year n,
  I = interest payment for year n,
  t = taxes for year n,
  P = principal payment for year n.

Rewriting the equation yields

$$G = X_n + C + I + t + P.$$

The above equation assumes that there are no capital requirements, salvage value considerations, or working capital changes in year n. For the minimum annual gross income, the cash flow, X_n, must satisfy the following relationship:

$$X_n = D_e + f_n,$$

where
   D_e = recovered portion of the equity capital
   f_n = return on the unrecovered equity capital.

It is assumed that the total equity and debt capital recovered in a year are equal to the book depreciation, D_b, and that the principal payments are a constant percentage of the book depreciation. That is,

$$P = c(D_b),$$

where c is the debt ratio. The recovery of equity capital is, therefore, given by the following:

$$D_e = (1-c)D_b.$$

The annual returns on equity, f_n, and interest, I, are based on the unrecovered balance as shown below:

$$f_n = (1-c)k_e(\mathrm{BV}_{n-1})$$

$$I = ck_d(\mathrm{BV}_{n-1})$$

where
   c = debt ratio
   k_e = required rate of return on equity
   k_d = required rate of return on debt capital
   BV_{n-1} = book value at the beginning of year n.

Based on the preceding equations, the minimum annual gross income, or revenue requirement, for year n can be represented as

$$R = D_b + f_n + C + I + t.$$

An expression for taxes, t, is given by

$$t = (G - C - D_t - I)T,$$

where
   D_t = depreciation for tax purposes
   T = tax rate.

If the expression for R is substituted for G in the above equation, the following alternate expression for t can be obtained:

$$t = \left[T/(1-T) \right] (D_b + f_n - D_t).$$

The calculated minimum annual revenue requirement can be used to evaluate the economic feasibility of a project. An example of a decision criterion that may be used for that purpose is presented below:

Decision Criterion: If expected gross incomes are greater than the minimum annual revenue requirements, then the project is considered to be economically acceptable and the project investment is considered to be potentially profitable. Economic acceptance should be differentiated from technical acceptance, however. If, of course, other alternatives being considered have similar results, a comparison based on the margin of difference (i.e., incremental analysis) between the expected gross incomes and minimum annual requirements must be made. There are two extensions to the basic analysis procedure presented above. They are the *flow-through* and *normalizing methods.*

This extension of the basic revenue requirement analysis allocates credits and costs in the year that they occur. That is, there are no deferred taxes and the investment tax credit is not amortized. Capitalized interest is taken as an expense in the first year. The resulting equation for calculating the minimum annual revenue requirements is

$$R = D_b + f_e + I + gP + C + t,$$

where the required return on equity is given by the following:

$$f_e = k_e (1-c) K_{n-1},$$

where
k_e = implied cost of common stock
c = debt ratio
K_{n-1} = chargeable investment for the preceding year
$K_n = K_{n-1} - D_b$ (with K_0 = initial investment)
g = capitalized interest rate.

The capitalized interest rate is usually set by federal regulations. The debt interest is given by

$$I = (c) k_d K_{n-1},$$

where
k_d = after-tax cost of capital.

The investment tax credit is calculated as follows:

$$C_t = i_t P,$$

where i, is the investment tax credit. Costs, C, are estimated totals that include such items as ad valorem taxes, insurance costs, operation costs, and maintenance costs. The taxes for the flow-through method are calculated as

$$t = \frac{T}{1-T}(f_e + D_b - D_i) - \frac{C}{1-T}.$$

The normalizing method differs from the flow-through method, in that deferred taxes are utilized. These deferred taxes are sometimes included as expenses in the early years of the project and then as credits in later years. This *normalized* treatment of the deferred taxes is often used by public utilities to minimize the potential risk of changes in tax rules that may occur before the end of the project but are unforeseen at the start of the project. Also, the interest paid on the initial investment cost is capitalized. That is, it is taken as a tax deduction in the first year of the project and then amortized over the life of the project to spread out the interest costs. The resulting minimum annual revenue requirement is expressed as

$$R = D_b + d_t + C_t - A_t + I + f_e + t + C,$$

where the depreciation schedules are based on the following capitalized investment cost:

$$K = P + gP,$$

with P and g as previously defined. The deferred taxes, d, are the difference in taxes that result from using an accelerated depreciation model instead of a straight line rate over the life of the project. That is,

$$d_t = (D_t - D_s)T,$$

where
D_t = accelerated depreciation for tax purposes
D_s = straight line depreciation for tax purposes.

The amortized investment tax credit, A_t, is spread over the life of the project, n, and is calculated as follows:

$$A_t = \frac{C_t}{n}.$$

The debt interest is similar to the earlier equation for capitalized interest. However, the chargeable investment differs by taking into account the investment tax credit, deferred taxes, and the amortized investment tax credit. The resulting expressions are

$$I = k_d(c) K_{n\,1},$$

$$K_n = K_{n-1} - D_b - C_t - d_t - A_t.$$

In this case, the expression for taxes, t, is given by the following:

$$t = \frac{T}{1-T}(f_e + D_b + d_t + C_t - A_t - D_t - gP) - \frac{C_t}{1-T}.$$

The differences between the procedures for calculating the minimum annual revenue requirements for the flow-through and the normalizing methods yield some interesting and important details. If the MARRs are converted to uniform annual amounts (leveled), a better comparison between the effects of the calculations for each method can be made. For example, the MARR data calculated by using each method are presented in Table 4.3.

The annual MARR values are denoted by R_n, and the uniform annual amounts are denoted by R_u. The uniform amounts are found by calculating the present value for each early amount and then converting that total amount to equal yearly amounts over the same span of time. For a given investment, the flow-through method will produce a smaller leveled minimum annual revenue requirement. This is because the normalized data include an amortized investment tax credit as well as deferred taxes. The yearly data for the flow-through method should give values closer to the actual cash flows, because credits and costs are assigned in the year in which they occur and not up front, as in the normalizing method.

The normalizing method, however, provides for a faster recovery of the project investment. For this reason, this method is often used by public utility companies when establishing utility rates. The normalizing method also agrees better, in

TABLE 4.3
Normalizing Versus Flow-Through Revenue Analysis

Year	Normalizing		Flow-Through	
	R_n	R_u	R_n	R_u
1	7,135	5,661	5,384	5,622
2	6,433	5,661	6,089	5,622
3	5,840	5,661	5,913	5,622
4	5,297	5,661	5,739	5,622
5	4,812	5,661	5,565	5,622
6	4,380	5,661	5,390	5,622
7	4,005	5,661	5,214	5,622
8	3,685	5,661	5,040	5,622

practice, with the required accounting procedures used by utility companies than the flow-through method does. Return on equity also differs between the two methods. For a given internal rate of return, the normalizing method will give a higher rate of return on equity than will the flow-through method. This difference occurs because of the inclusion of deferred taxes in the normalizing method. Suppose we have the following data for a project. It is desired to perform a revenue requirement analysis using both the flow-through and the normalizing methods.

Initial Project Cost	= $100,000
Book Salvage Value	= $10,000
Tax Salvage Value	= $10,000
Book Depreciation Model	= Straight Line
Tax depreciation Model	= Sum-of-Years Digits
Life for Book Purposes	= 10 years
Life for Tax Purposes	= 10 years
Total Costs per Year	= $4,000
Debt Ratio	= 40%
Required Return of Equity	= 20%
Required Return on Debt	= 10%
Tax Rate	= 52%
Capitalized Interest	= 0%
Investment Tax Credit	= 0%

Tables 4.4–4.7 show the differences between the normalizing and flow-through methods for the same set of data. The different treatments of capital investment produced by the investment tax credit can be seen in the tables as well as in Figure 4.1–Figure 4.5.

TABLE 4.4
Part One of MARR Analysis

	Tax Depreciation		Deferred Taxes	
Year	Normalizing	Flow-Through	Normalizing	Flow-Through
1	16363.64	16363.64	3829.09	None
2	14727.27	14727.27	2978.18	
3	13090.91	13090.91	2127.27	
4	11454.55	11454.55	1276.36	
5	9818.18	9818.18	425.45	
6	8181.82	8181.82	−425.45	
7	6545.45	6545.45	−1276.36	
8	4909.09	4909.09	−2127.27	
9	3272.73	3272.73	−2978.18	
10	1636.36	1636.36	−3829.09	

TABLE 4.5
Part Two of MARR Analysis

	Capitalized Investment		Taxes	
Year	Normalizing	Flow-Through	Normalizing	Flow-Through
	100000.00	100000.00	—	—
1	87170.91	91000.00	9170.91	5022.73
2	75192.73	92000.00	8354.04	5625.46
3	64065.46	73000.00	7647.78	6228.18
4	53789.90	64000.00	7052.15	6830.91
5	44363.64	55000.00	6567.13	7433.64
6	35789.09	46000.00	6192.73	8036.36
7	28065.45	37000.00	5928.94	8639.09
8	21192.72	28000.00	5775.78	9241.82
9	15170.90	19000.00	5733.24	9844.55
10	10000.00	10000.00	5801.31	10447.27

TABLE 4.6
Part Three of MARR Analysis

	Return on Debt		Return of Equity	
Year	Normalizing	Flow-Through	Normalizing	Flow-Through
1	4000.00	4000.00	12000.00	12000.00
2	3486.84	3640.00	10460.51	10920.00
3	3007.71	3280.00	9023.13	9840.00
4	2562.62	2920.00	7687.86	8760.00
5	2151.56	2560.00	6454.69	7680.00
6	1774.55	2200.00	5323.64	6600.00
7	1431.56	1840.00	4294.69	5520.00
8	1122.62	1480.00	3367.85	4440.00
9	847.71	1120.00	2543.13	3360.00
10	606.84	760.00	1820.51	2280.00

There is a big difference in the distribution of taxes, since most of the taxes are paid early in the investment period with the normalizing method, but taxes are deferred with the flow-through method. The resulting minimum annual revenue requirements are larger for the normalizing method early in the period. However, there is a more gradual decrease with the flow-through method. Therefore, the use of the flow-through method does not put as great a demand on the project to produce high revenues early in the project life cycle as does the normalizing method.

TABLE 4.7
Part Four of MARR Analysis

	Minimum Annual Revenues	
Year	Normalizing	Flow-Through
1	42000.00	34,022.73
2	38279.56	33185.45
3	34805.89	32348.18
4	31578.98	31510.91
5	28598.84	30673.64
6	25865.45	29836.36
7	23378.84	2899.09
8	21138.98	28161.82
9	19145.89	27324.55
10	17,399.56	26487.27

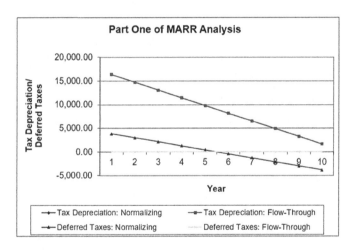

FIGURE 4.2 Plot of part one of MARR analysis.

Also, the normalizing method produces a lower rate of return on equity. This fact may be of particular interest to shareholders.

This chapter has presented selected general techniques of applied economic analysis for industrial projects, suitable for utilization in the evaluation stage of DEJI Systems Model. Applied economic analysis techniques, as presented in this chapter, are useful for evaluating system-based projects (designs and options) in engineering, industry, business, government, academia, the military, and other enterprises.

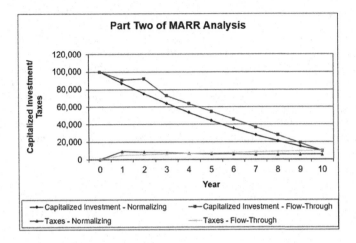

FIGURE 4.3   Plot of part two of MARR analysis.

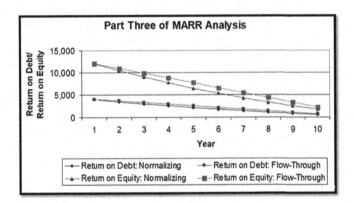

FIGURE 4.4   Plot of part three of MARR analysis.

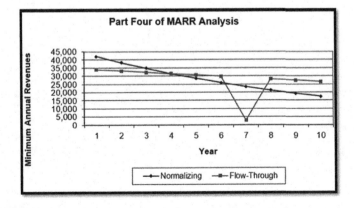

FIGURE 4.5   Plot of part four of MARR analysis.

REFERENCES

Badiru, A. B., and O. A. Omitaomu (2007), *Computational Economic Analysis for Engineering and Industry*, Taylor & Francis Group/CRC Press, Boca Raton, FL.

Badiru, A. B., and David Russell (1987), "Minimum Annual Project Revenue Requirement Analysis", *Proceedings of 9th Annual Conference on Computers and Industrial Engineering*, Atlanta, Georgia, March, *Computers and Industrial Engineering*, Vol. 13, Nos. 1–4, 1987, pp. 366–370.

Badiru, A. B., and D. B. Sieger, (1998) "Neural Network as a Simulation Metamodel in Economic Analysis of Risky Projects," *European Journal of Operational Research*, Vol. 105, pp. 130–142.

5 Justification in DEJI Systems Model®

Not everything that is feasible is practical and desirable.

—Adedeji Badiru

INTRODUCTION

The above quote sums up the essence of why the justification stage of DEJI Systems Model® is needed. A design that evaluates well may not be justified for implementation for various reasons. The question at the root of a systems design centers around what is justified, what is allowed, or what is possible.

Sometimes, there is a fine line between system cost and system value because the metrics can be used interchangeably in evaluating and justifying a system. While cost may be used as the primary basis for evaluating a system, value may be the basis for justifying the system, under the structure of DEJI Systems Model. Even something that has no cost may have a high value for the purpose of justification. Conversely, a high cost does not, necessarily, translate to a high value. Figure 5.1 conveys this graphically.

VALUE "ILITIES" OF A SYSTEM

It is at this stage of DEJI Systems Model that the many "ilities" of a system may come into play. Some of such "ilities" and "lities" are presented below:

• Ability	• Fragility	• Serviceability
• Affordability	• Habitability	• Suitability
• Agility	• Interchangeability	• Survivability
• Applicability	• Maintainability	• Testability
• Capability	• Modularity	• Transmittability
• Configurability	• Practicality	• Usability
• Desirability	• Predictability	• Utility
• Desirability	• Quality	• Variability
• Fidelity	• Reachability	• Vulnerability
• Flexibility	• Reliability	• Workability

For example, survivability refers to the consideration of system-design features that reduce the risk of demise from the system's operation, in the event of a catastrophic development. For example, airbags and seatbelts increase the survivability

DOI: 10.1201/9781003175797-5

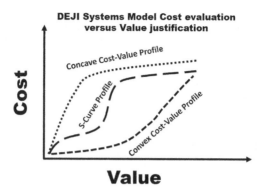

FIGURE 5.1 Cost evaluation versus value justification in DEJI Systems Model.

potential following an auto accident. Similarly, agility has the attributes of flexibility, adaptability, responsiveness, robustness, and innovativeness. In defense acquisition organizations, much research, both quantitative and qualitative, has been directed at measuring agility with respect to the interplay of human, social, technical, and cyber subsystems. This chapter presents the application of utility modeling for the assessment of the value and applicability of a system, particularly if some sort of innovation is involved. Innovation can be useful on varying levels and to differing extents. Utility modeling provides quantitative insights into the usefulness and desirability of innovation. Along with the quantitative assessment, an organization can also incorporate qualitative assessment along the lines of some of the ilities identified above.

SYSTEMS UTILITY MODELING

For the purpose of a contextual focus, the discussions that follow will use "systems" and "innovation" interchangeably because every system contains some aspect of innovation. Conversely, every innovation is embedded in a system. We, by default, believe that every system has utility to the same desirable extent. Unfortunately, it is not always the case. Every system has utility on differing scales and to differing extents. In fact, for social reasons, some innovations may not be desirable at all. For example, in warfare history, the invention of the Gatling Gun was hailed by some while abhorred by others. So, innovation can mean different things to different groups. For this reason, the technique of utility modeling is applicable for the assessment of innovation. The advantage of using utility modeling for innovation is that assessment metrics are imposed on the expected performance of innovation. How do we know if and when innovation has occurred and to what extent? It is easy to claim to be innovative or to proclaim commitment to innovation, but it is a different thing to be able to present a quantifiable proof of innovation. Innovation investment selection is an important aspect of investment planning. The right investment must be undertaken at the right time to satisfy the constraints of time and resources. A combination of criteria can be used to help in investment selection, including technical merit, management desire, schedule efficiency, cost-benefit ratio, resource availability, criticality of need, availability of sponsors, and user acceptance.

Many aspects of investment selection cannot be expressed in quantitative terms. For this reason, investment analysis and selection must be addressed by techniques that permit the incorporation of both quantitative and qualitative factors. Some techniques for investment analysis and selection are presented in the sections that follow. These techniques facilitate the coupling of quantitative and qualitative considerations in the investment decision process. Such techniques as net present value, profit ratio, and equity break-even point, which have been presented in the preceding sections, are also useful for investment selection strategies.

QUANTIFYING UTILITY

The term "utility" refers to the rational behavior of a decision-maker faced with making a choice in an uncertain situation. The overall utility of a system can be measured in terms of both quantitative and qualitative factors. This section presents an approach to system investment assessment based on utility models that have been developed within an extensive body of literature. The approach fits an empirical utility function to each factor that is to be included in a multi-attribute selection model. The specific utility values (weights) that are obtained from the utility functions are used as the basis for selecting a system.

Utility theory is a branch of decision analysis that involves the building of mathematical models to describe the behavior of a decision-maker faced with making a choice among alternatives in the presence of risk. Several utility models are available in the management science literature. The utility of a composite set of outcomes of n decision factors is expressed in the general form below:

$$U\left(x\right) = U\left(x_1, x_2, \ldots x_n\right)$$

where x_i = specific outcome of attribute X_i, $i = 1, 2, \ldots, n$ and $U(x)$ is the utility of the set of outcomes to the decision-maker. The basic assumption of utility theory is that people make decisions with the objective of maximizing those decisions' *expected utility*. Drawing on an example presented by Park and Sharp-Bette (1990), we may consider a decision-maker whose utility function with respect to system selection is represented by the following expression:

$$u\left(x\right) = 1 - e^{-0.0001}, $$

where x represents a measure of the benefit derived from a system. Benefit, in this sense, may be a combination of several factors (e.g., quality improvement, cost reduction, or productivity improvement) that can be represented in dollar terms.

Suppose this decision-maker is faced with a choice between two system alternatives, each of which has benefits as specified below:

System 1: Probabilistic levels of system benefits.

Benefits (x): −$10,000, $0, $10,000, $20,000, $30,000
Probabilities ($P(x)$): 0.2, 0.2, 0.2, 0.2, 0.2

System 2: A definite benefit of $5,000.

Assuming an initial benefit of zero and identical levels of required investment, the decision-maker must choose between the two systems. For System 1, the expected utility is computed as shown below:

$$E\left[u\left(x\right)\right] = \Sigma u\left(x\right)\left\{P\left(x\right)\right\},$$

For System 1, using the utility function calculations and the probability values provided, we have the following series of calculations for the expected utility:

−$10,000(−1.7183)(0.2) = −0.3437
$0(0, 0.2) = 0
$10,000(0.6321)(0.2) = 0.1264
$20,000(0.8647)(0.2) = 0.1729
$30,000(0.9502)(0.2) = 0.1900

Thus, $E[u(x)_1] = 0.1456$.

For System 2, we have $u(x)_2 = u(\$5000) = 0.3935$. Consequently, the system providing the certain amount of $5000 is preferred to the riskier System 1 investment, even though System 1 has a higher expected benefit of $\Sigma xP(x) = \$10,000$. A plot of the utility function used in the above example is presented in Figure 5.2. In general, the profile of a utility curve can be convex, concave, S-curved, or some compound curvature stretching from left to right.

If the expected utility of 0.1456 is set equal to the decision-maker's utility function, we obtain the following:

$$0.1456 = 1 - e^{-0.0001x*},$$

which yields $x* = \$1,574$, referred to as the *certainty equivalent* (CE) of System I ($CE_1 = 1574$). The certainty equivalent of an alternative with variable outcomes is a *certain amount* (CA), which a decision-maker will consider to be desirable to the same degree as the variable outcomes of the alternative. In general, if CA represents the certain amount of benefit that can be obtained from System II, then the criteria for making a choice between the two systems can be summarized as follows:

If CA < $1,574, select System I
If CA = $1,574, select either system
If CA > $1,574, select System II.

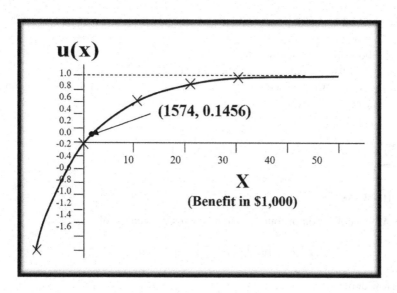

FIGURE 5.2   Example of utility function and certainty equivalent.

The key to using utility theory for a system justification selection is choosing the proper utility model. The sections that follow describe two widely used utility models:

- Additive utility model
- Multiplicative utility model

ADDITIVE UTILITY FUNCTION

The additive utility of a combination of outcomes of n factors $(X_1, X_2, ..., X_n)$ is expressed as follows:

$$U(x) = \sum_{i=1}^{n} U\left(x_i, \bar{x}_i^0\right)$$

$$= \sum_{i=1}^{n} k_i U(x_i),$$

where
   x_i = measured or observed outcome of attribute i
   n = number of factors to be compared
   x = combination of the outcomes of n factors
   $U(x_i)$ = utility of the outcome for attribute i, x_i
   $U(x)$ = combined utility of the set of outcomes, x
   k_i = weight or scaling factor for attribute $i(0 < k_i < 1)$

X_i = variable notation for attribute i

x_i^0 = worst outcome of attribute i

x_i^* = best outcome of attribute i

$\overline{x}_i^0$ = set of worst outcomes for the complement of x_i

$U\left(x_i, \overline{x}_i^0\right)$ = utility of the outcome of attribute I and the set of worst outcomes
for the complement of attribute i

$k_i = U\left(x_i^*, \overline{x}_i^0\right)$

$\sum_{i=1}^{n} k_i = 1.0$ (required for the additive model).

UTILITY EXAMPLE

Let **A** be a collection of four system attributes defined as follows:

$$\mathbf{A} = \{\text{Profit,Flexibility,Quality,Productivity}\}.$$

Now define:

$$\mathbf{X} = \{\text{Profit,Flexibility}\} \text{ as a subset of A.}$$

Then, $\overline{\mathbf{X}}$ is the complement of **X** defined as follows:

$$\overline{\mathbf{X}} = \{\text{Quality,Productivity}\}$$

An example of the comparison of two investments under the additive utility model is summarized in Table 5.1 and yields the following results:

$$U(x)_A = \sum_{i=1}^{n} k_i U_i(x_i) = .4(.95) + .2(.45) + .3(.35) + .1(.75) = 0.650$$

$$U(x)_B = \sum_{i=1}^{n} k_i U_i(x_i) = .4(.90) + .2(.98) + .3(.20) + .1(.10) = 0.626$$

since $U(x)_A > U(x)_B$, Investment A is selected.

TABLE 5.1
Example of Additive Utility Computation

Attribute, i	Weight, k_i	Investment A $U_i(x_i)$	Investment B $U_i(x_i)$
Profitability	0.4	0.95	0.90
Flexibility	0.2	0.45	0.98
Quality	0.3	0.35	0.20
Throughput	0.1	0.75	0.10

MULTIPLICATIVE UTILITY MODEL

Under the multiplicative utility model, the utility of a combination of outcomes of n factors $(X_1, X_2, \ldots, X_{n1})$ is expressed as

$$U(x) = \frac{1}{C}\left[\prod_{i=1}^{n}\left(Ck_iU_i(x_i)+1\right)-1\right],$$

where C and k_i are scaling constants satisfying the following conditions:

$$\prod_{i=1}^{n}(1+Ck_i)-C = 1.0$$

where $-1.0 < C < 0.0$ and $0 < k_i < 1$.

The other variables are as defined previously for the additive model. Using the multiplicative model for the data in Table 5.1 yields $U(x)_A = 0.682$ and $U(x)_B = 0.676$. Thus, System A is the better option.

FITTING A UTILITY FUNCTION

An approach presented in this section for multi-attribute investment selection is to fit an empirical utility function to each factor to be considered in the selection process. The specific utility values (weights) that are obtained from the utility functions may then be used in any of the standard investment justification methodologies. One way to develop empirical utility function for an investment attribute is to plot the "best" and "worst" outcomes expected from the attribute and then to fit a reasonable approximation of the utility function using concave, convex, linear, S-shaped, or any other logical functional form.

Alternately, if an appropriate probability density function can be assumed for the outcomes of the attribute, then the associated cumulative distribution function may yield a reasonable approximation of the utility values between 0 and 1 for corresponding outcomes of the attribute. In that case, the cumulative distribution function gives an estimate of the cumulative utility associated with increasing levels of attribute outcome. Simulation experiments, histogram plotting, and goodness-of-fit tests may be used to determine the most appropriate density function for the outcomes of a given attribute. For example, the following five attributes are used to illustrate how utility values may be developed for a set of system attributes. The attributes are return on investment (ROI), productivity improvement, quality improvement, idle time reduction, and safety improvement.

EXAMPLE

Suppose we have historical data on the return on investment (ROI) for investing in a particular investment. Assume that the recorded ROI values range from 0% to 40%. Thus, the worst outcome is 0% and the best outcome is 40%. A frequency distribution of the observed ROI values is developed and an appropriate probability density function (pdf) is fitted to the data. For our example, suppose the ROI is found to be exceptionally distributed with a mean of 12.1%. That is:

$$f(x) = \begin{cases} \dfrac{1}{\beta} e^{-x/\beta}, & \text{if } x \geq 0 \\ 0, & \text{otherwise} \end{cases}$$

$$F(x) = \begin{cases} 1 - e^{-x/\beta}, & \text{if } x \geq 0 \\ 0, & \text{otherwise} \end{cases}$$

$$\approx U(x),$$

where $\beta = 12.1$. $F(x)$ approximates $U(x)$. The probability density function and cumulative distribution function can be plotted graphically. The utility of any observed ROI within the applicable range may be read directly from the plotted cumulative distribution function. For the productivity improvement attribute, suppose it is found (based on historical data analysis) that the level of improvement is normally distributed with a mean of 10% and a standard deviation of 5%. That is,

$$f(x) = \frac{1}{\sqrt{2\pi}\sigma} e^{-\frac{1}{2}\left(\frac{x-\mu}{\sigma}\right)^2}, \quad -\infty < x < \infty$$

where $\pi = 10$ and $\sigma = 5$. Since the normal distribution does not have a closed-form expression for $F(x)$, $U(x)$ is estimated by plotting representative values based on the standard normal table. A plot can be generated to show $f(x)$ and the estimated utility function for productivity improvement. The utility of productivity improvement may also be evaluated on the basis of cost reduction. Suppose quality improvement is subjectively assumed to follow a beta distribution with shape parameters $\alpha = 1.0$ and $\beta = 2.9$. That is,

$$f(x) = \frac{\Gamma(\alpha + \beta)}{\Gamma(\alpha)\Gamma(\beta)} \cdot \frac{1}{(b-a)^{\alpha+\beta-1}} \cdot (x-a)^{\alpha-1} (b-x)^{\beta-1},$$

for $a \leq x \leq b$ and $\alpha > 0$, $\beta > 0$.

where a = lower limit for the distribution
b = upper limit for the distribution
α, β are the shape parameters for the distribution.

As with the normal distribution, there is no closed-form expression for $F(x)$ for the beta distribution. However, if either of the shape parameters is a positive integer, then a binomial expansion can be used to obtain $F(x)$. Based on work analysis observations, suppose idle time reduction is found to be best described by a log-normal distribution with a mean of 10% and standard deviation of 5%. This is represented as shown below:

$$f(x) = \frac{1}{x\sqrt{2\pi\sigma^2}} e^{\left[\frac{-(\ln x - \mu)^2}{2\sigma^2}\right]}, \quad x > 0.$$

There is no closed-form expression for $F(x)$. For example, suppose safety improvement is assumed to have a previously known utility function, defined as follows:

$$U_p(x) = 30 - \sqrt{400 - x^2},$$

where x represents percent improvement in safety. For the expression, the un-scaled utility values range from 10 (for 0% improvement) to 30 (for 20% improvement). To express any particular outcome of an attribute i, x_i, on a scale of 0.0 to 1.0, it is expressed as a proportion of the range of best to worst outcomes shown below:

$$X = \frac{x_i - x_i^0}{x_i^* - x_i^0}$$

where X = outcome expressed on a scale of 0.0 to 1.0
x_i = measured or observed raw outcome of attribute i
x_i^0 = worst raw outcome of attribute i
x_i^* = best raw outcome of attribute i.

The utility of the outcome may then be represented as $U(X)$ and read off of the empirical utility curve. Using the above approach, the utility function for safety improvement is scaled from 0.0 to 1.0. The respective utility values for the five attributes may be viewed as relative weights for comparing investment alternatives. The utility obtained from the modeled functions can be used in the additive and multiplicative utility models discussed earlier. For example, Table 5.2 shows a composite utility profile for a proposed innovation system.

TABLE 5.2
Composite Utility for a Proposed System

Attribute, i	k_i	Value	$U_i(x_i)$
Return on investment	0.30	12.1%	0.61
Productivity improvement	0.20	10.0%	0.49
Quality improvement	0.25	60.0%	0.93
Idle time reduction	0.15	15.0%	0.86
Safety improvement	0.10	15.0%	0.40

Using the additive utility model, the *composite utility* (CU) of the investment, based on the five attributes, is given by:

$$U(X) = \sum_{i=1}^{n} k_i U_i(x_i) = .30(.61) + .20(.49) + .25(.93) + .15(.86) + .10(.40) = 0.6825.$$

This composite utility value may then be compared with the utilities of other investments. On the other hand, a single investment may be evaluated independently on the basis of some minimum acceptable level of utility (MALU) desired by the decision-maker. The criteria for evaluating an investment based on MALU may be expressed by the following rule:

Investment j is acceptable if its composite utility, $U(X)_j$, is greater than MALU; Investment j is not acceptable if its composite utility, $U(X)_j$, is less than MALU.

The utility of an investment may be evaluated on the basis of its economic, operational, or strategic importance to an organization. Utility functions can be incorporated into existing justification methodologies. For example, in the analytic hierarchy process, utility functions can be used to generate values that are, in turn, used to evaluate the relative preference levels of attributes and alternatives. Utility functions can be used to derive component weights when the overall effectiveness of investments is being compared. Utility functions can generate descriptive levels of investment performance as well as indicating the limits of investment effectiveness.

COMPOSITE VALUE COMPUTATION

A technique that is related to the utility modeling is the investment value model (PVM), which is an adaptation of the manufacturing system value (MSV) model presented by Troxler and Blank (1989). The model is suitable for the incorporation of utility values, and an example is presented below. The model provides a heuristic decision aid for comparing investment alternatives. "Value" is represented as a determined vector function that indicates the value of tangible and intangible "attributes" that characterize an alternative. Value can be expressed as follows:

$$V = f\left(A_1, A_2 \ldots A_p\right)$$

where V = value, $A = (A_1,\ldots,A_n)$ = vector of quantitative measures or attributes, and p = the number of attributes that characterize the investment. Examples of investment attributes include quality, throughput, capability, productivity, and cost performance. Attributes are considered to be a combined function of "factors," x_i, expressed as

$$A_k\left(x_1, x_2, \ldots, x_{m_k}\right) = \sum_{i=1}^{m_k} f_i\left(x_i\right)$$

where $\{x_i\}$ = set of m factors associated with attribute A_k ($k = 1,2,\ldots,p$) and f_i = contribution function of factor x_i to attribute A_k. Examples of factors are market share, reliability, flexibility, user acceptance, capacity utilization, safety, and design functionality. Factors are themselves considered to be composed of "indicators," v_i, expressed as

$$x_i\left(v_1, v_2, \ldots, v_n\right) = \sum_{j=1}^{n} z_i\left(v_i\right)$$

where $\{v_j\}$ = set of n indicators associated with factor x_i ($i = 1,2,\ldots,m$) and z_j = scaling function for each indicator variable v_j. Examples of indicators are debt ratio, investment responsiveness, lead time, learning curve, and scrap volume. By combining the above definitions, a composite measure of the value of an investment is given by the following:

$$PV = f\left(A_1, A_2, \ldots, A_p\right) =$$

$$f\left\{\left[\sum_{i=1}^{m_1} f_i\left(\sum_{j=1}^{n} z_j\left(v_j\right)\right)\right]_1, \left[\sum_{i=1}^{m_2} f_i\left(\sum_{j=1}^{n} z_j\left(v_j\right)\right)\right]_2, \ldots, \left[\sum_{i=1}^{m_k} f_i\left(\sum_{j=1}^{n} z_j\left(v_j\right)\right)\right]_p\right\}$$

where m and n may assume different values for each attribute. A subjective measure to indicate the utility of the decision-maker may be included in the model by using an attribute weighting factor, w_i, to obtain the following:

$$PV = f\left(w_1 A_1, w_2 A_2, \ldots w_p A_p\right)$$

TABLE 5.3
Comparison of Values of Alternate Systems

System Alternatives	Suitability $k = 1$	Capability $k = 2$	Performance $k = 3$	Productivity $K = 4$
System A	0.12	0.38	0.18	0.02
System B	0.30	0.40	0.28	1.00
System C	0.53	0.33	0.52	1.10

where

$$\sum_{k=1}^{p} w_k = 1, \quad \left(0 \le w_k \le 1\right).$$

As an example, an analysis using the above model to compare three investments on the basis of four attributes is presented in Table 5.3. The four attributes, *capability, suitability, performance,* and *productivity*, require careful interpretation before relative weights for the alternatives can be developed.

CAPABILITY

The term "capability" refers to the ability of equipment to produce certain features. For example, a certain piece of equipment may only produce horizontal or vertical slots, and flat finishes. But a multi-axis machine can produce spiral grooves, internal metal removal from prismatic or rotational parts, thus, increasing the part variety that can be made. In Table 5.3, the levels of increase in part variety from the three competing systems are 38%, 40%, and 33%, respectively.

SUITABILITY

"Suitability" refers to the appropriateness of the system to company operations. For example, chemical milling is more suitable for making holes in thin, flat metal sheets than drills are. Drills need special fixtures to hold the thin metal down and protect it from wrinkling and buckling. The parts that the three systems are suitable for are respectively, 12%, 30%, and 53% of the current part mix.

PERFORMANCE

"Performance" in this context refers to the ability of the system to produce high-quality outputs, or the ability to meet extra-tight performance requirements. In our example, the three systems can, respectively, meet tightened standards on 18%, 28%, and 52% of the normal set of jobs.

Productivity

"Productivity" can be measured by a simulation of the performance of the current system with the proposed technology at its current production rate, quality level, and part application. For example, in Table 5.3, the three systems, respectively, show increases of 0.02, −1.0, and −1.1 on a uniform scale of productivity measurement.

A plot of the histograms of the respective "values" of the three systems shows that System C is the best alternative in terms of suitability and performance. System B shows the best capability measure, but its productivity is too low to justify the needed investment. System A, then, offers the best productivity, but its suitability measure is low. The relative weights used in many justification methodologies are based on subjective propositions of the decision-maker(s). Some of those subjective weights can be enhanced by the incorporation of utility models. For example, the weights shown in Table 5.3 could be obtained from utility functions.

VALUE BENCHMARKING FOR SYSTEM JUSTIFICATION

The techniques presented in the preceding sections can be used for benchmarking with respect to desirable values of competing systems. For example, to develop a baseline assessment, evidence of successful practices from other systems may be needed to justify a system, regardless of how well it performs at the evaluation stage of DEJI Systems Model. Metrics based on an organization's most critical system implementation issues should be developed. Benchmarking is a process whereby target performance standards are established based on the best examples available. The objective is to equal or surpass the best example. In its simplest term, benchmarking means learning from and emulating a superior example. The premise of benchmarking is that if an organization replicates the best examples, it will become one of the best in the business. A major approach of benchmarking is to identify performance gaps between systems. Benchmarking requires that an attempt be made to close the gap by improving the performance of the system under consideration.

Benchmarking requires frequent comparison with the target system. Updates must be obtained from the systems that are benchmarked. The new system to be benchmarked must be selected on a periodic basis. Measurement, analysis, feedback, and modification should be incorporated into the performance improvement program. The benchmark-feedback model is useful for establishing a continuous drive toward performance benchmarks. A graph can be used to represent the input-output relationships of the elements in a benchmarking environment. The inputs may be in terms of data, information, raw material, technical skill, or other basic resources that work together to impart value to the system of interest. A feedback process is necessary to determine what control actions should be taken at the next improvement phase. The primary responsibility of an system analyst is to ensure proper forward and backward flow of information concerning the performance of a system on the basis of the benchmarked targets.

CONCLUDING REMARKS ON JUSTIFICATION

The premise of this chapter is that a system that has gone through the design and evaluation stages of the DEJI Systems Model should be justified on the basis of a combination of quantitative and qualitative parameters. The quantitative methods illustrated in this chapter are just a limited selection of the wide range of techniques available. The point is that readers should leverage whatever soft and hard tools are available to pass a system through all the stages of DEJI Systems Model, as summarized again below:

Design
Evaluation
Justification
Integration

On the quantitative side, some of the system cost parameters that may be encountered in the application of DEJI Systems Model include the following:

Life Cycle Cost: Sum of all the associated costs over the life of an innovation project
First Cost: Total initial investment for an innovation project
Operating Cost: Recurring cost needed to operate innovation
Maintenance Cost: Recurring cost needed to maintain innovation in a good working condition
Sunk Cost: Past cost in innovation that cannot be reversed
Opportunity Cost: Cost of forgoing the opportunity to earn an interest in innovation
Equity Capital: Actual innovation-driven cash on hand against which there are no debt obligations
Direct Cost: Costs that are directly associated with actual operation of innovation
Indirect Cost: Cost that is indirectly related to operation of innovation
Overhead Cost: Cost incurred for activities performed in support of the system
MARR: Minimum acceptable rate of return
Cost of Capital: Cost of obtaining capital funds
Fixed Cost: Cost incurred irrespective of the level of system implementation
Variable Cost: Cost that varies in direct proportion to the level of system implementation
Marginal Cost: Cost of increasing production level by one additional unit in the system
Incremental Cost: Cost of advancing the system from one level to another
Disposal Cost: Cost associated with getting rid of a system post implementation
Salvage Cost: Terminal value of a system at the end of its service life
Scrap Value: Residual value of a system

Future Cost: Expected cost of owning or retaining the system beyond the present time

Economies of Scale: Effect of mass production due to the system (reduction of the relative weight of the fixed cost in the total cost by increasing the output quantity)

For the non-manufacturing environment, the types of costs encountered in a system may be different from the examples presented above. Additional tools and techniques applicable for the justification Stage of DEJI Systems Model can be found in Badiru (2012), Badiru (1991), Badiru (2008), Jaber (2011), Park and Sharp-Bette (1990), Saaty (1980), and Troxler and Blank (1989).

REFERENCES

Badiru, A. B. (2012), *Project Management: Systems, Principles, and Applications*, Taylor & Francis Group/CRC Press, Boca Raton, FL.

Badiru, Adedeji B. (1991), *Project Management Tools for Engineering and Management Professionals*, Industrial Engineering & Management Press, Norcross, GA.

Badiru, Adedeji B. (2008), *Triple C Model of Project Management: Communication, Cooperation, and Coordination*, Taylor & Francis Group/CRC Press, Boca Raton, FL.

Jaber, M. (2011), *Learning Curves: Theory, Models, and Applications, editor*, CRC Press, Boca Raton, FL.

Park, Chan S., and Gunter P. Sharp-Bette (1990), *Advanced Engineering Economics*, Wiley & Sons, NY, 1990.

Saaty, Thomas L. (1980), *The Analytic Hierarchy Process*, McGraw-Hill, New York.

Troxler, Joel W., and Leland Blank (1989), "A Comprehensive Methodology for Manufacturing System Evaluation and Comparison," *Journal of Manufacturing Systems*, Vol. 8, No. 3, pp. 176–183.

6 Integration Stage of DEJI Systems Model®

> If there is no integration, there is no implementation.
>
> —Adedeji Badiru

INTRODUCTION

Integration is the basis for every system implementation (Badiru and Thomas, 2013). As the captioned quote of this chapter reminds us, "If there is no integration, there is no implementation." As a simple social rule, the author proclaims, "Happy-go-working people work better when they are happy." Integrate happiness into the work and see how better people work. Another pertinent quote to emphasize the general applicability of DEJI Systems Model® is provided below:

> You can sympathize from a distance, but you can empathize only from close range.
>
> —Adedeji Badiru

This means that in human interactions, a helping hand extended with a close-up relationship of integration is more effective than an offer of help lobbed from afar. To appreciate the importance of thinking of integration in the early stages of a system, we can consider the 2022 case of the nationwide implementation of 5G (fifth generation) technology for communication. The fifth generation of mobile networks (aka 5G) is the global standard for cellular mobile communications. The technology is based on the orthogonal frequency division multiplexing (OFDM) system. It succeeds the previous technology of 4G. It offers a completely different transmission spectrum under the wider bandwidth technologies. It leverages the technical capabilities of sub-6 GHz and mmWave. It not only improves connectivity but also permits a wider coverage of worldwide digital communications. The enhanced efficiency, effectiveness, and communication transmission quality have the benefits of handling high data rates, big data capability, reduced latency, energy saving, cost reduction, global connectivity, and higher system capacity. In essence, 5G satisfies many of the "ilities" of a system as described in the previous chapter. The first phase of 5G specifications in Release-15 was completed in 2019, to accommodate early commercial deployment. The second phase in Release-16 was completed in 2020. Implementation (deployment) was slated for 2022. With its proven technology, the worldwide desirability of 5G makes it highly justified

DOI: 10.1201/9781003175797-6

for implementation. Using the framework of DEJI Systems Model, we can summarize the flight and plight of 5G as follows:

Design: The technology of 5G has, no doubt, undergone a rigorous design process. The technical design is proven and the development has been theoretically and practically vetted through extensive scientific and engineering studies.

Evaluation: 5G has been extensively evaluated across diverse parameters and found to be worthy of what the world needs in a modern digital era.

Justification: 5G is in high demand and justified for implementation on the basis of its wide scope and global demand.

Integration: It is at the implementation stage that 5G faces its stiffest test because of an unforeseen integration issue with commercial air traffic. Just before the implementation date, 5G faced the following challenges:

- Major airline carriers in the United States warned that plans by mobile phone companies to use the communication spectrum for 5G wireless services could be highly disruptive to air flight communications in the vicinity of airports.
- The claim was that this could adversely affect pilots' communication with ground stations during landing approaches. 5G may adversely interfere with delicate and sensitive aviation instruments.
- A fear was expressed that the implementation could cost upward of $1.6 billion annually in air travel delays.

The above concerns exist despite the fact that 5G has been implemented successfully and safely in some countries. In recognition of the concerns, FAA (Federal Aviation Administration) issued new airworthiness directives, warning that interference from 5G wireless spectrum could result in flight diversions. This realization brings to mind the drag mentioned in William Shakespeare's quote: "things sweet to taste prove in digestion sour." If the sourness of implementation is not prepared for in advance, the implementation could be less than stellar. To be forewarned is to be fore-prepared. This is the essence of the integration stage of DEJI Systems Model. Meanwhile, AT&T and Verizon, major mobile phone companies, assert that their 5G equipment will not interfere with aircraft electronics. Only time and space will tell.

INTEGRATION BY OTHER NAMES

Integration, as contextualized in the DEJI Systems Model, can be viewed in different adaptive implications. In functionality, integration implies adaptation, assimilation, operational acceptance, alignment, acclimatization, adoption, absorption, inculcation, ingestion, alliance, orientation, affiliation, inclusivity, acculturation, accommodation, coordination, ingraining, agreement, coalition, resolution, ratification, rationalization, realism, pragmatism, orientation, compliance, conforming, correlating, and adjustment (as in employers adjusting to the workforce

TABLE 6.1
DEJI System Model for Technology Integration

DEJI Systems Model	Characteristics	Tools & Techniques
Design	Define goals	Parametric assessment
	Set performance metrics	Project state transition
	Identify milestones	Value stream analysis
Evaluate	Measure parameters	Pareto distribution
	Assess attributes	Life cycle analysis
	Benchmark results	Risk assessment
Justify	Assess economics	Benefit-cost ratio
	Assess technical output	Payback period
	Align with goals	Present value
Integrate	Embed in normal operation	SMART concept
	Verify symbiosis	Process improvement
	Leverage synergy	Quality control

crunch caused by the pandemic). In the context of a biological system, evolution is, indeed, a good example of integration with the prevailing environmental conditions. That is, indeed, the origin of Charles Darwin's "Origin of the Species," which conveys an adaptation (integration) to the local environment. Whatever an organization's interest, there is always a place for integration. The structured approach of DEJI Systems Model is to arrive at the integration stage after explicitly passing through the design, evaluation, and justification stages. Integration will be stronger and more sustainable if the preceding three stages are accomplished structurally. In many cases, we have to evaluate a system on the basis of independence versus reliance among the subsystems of the system. The essence of the DEJI Systems Model for technology integration is summarized in Table 6.1.

PRACTICAL MEASURING OF INTEGRATION

Depending on the nature of integration needed, measuring, ascertaining, or assessing the level of integration can be done quantitatively or qualitatively. If a qualitative assessment is used, informed subjective mechanisms may have to be used. An informed subjective mechanism simply means that the assessment is based on a direct observation or some experiential basis, beyond the typical seat-of-the-pant approaches. Case studies that are provided in the chapters that follow illustrate examples of implicit integration without resorting to esoteric quantitative modeling and computation. However, to show the full range of assessing integration, some selected quantitative measures are presented in this chapter.

It will not always be possible to model the parameters of subsystems in a system for the purpose of quantitative alignment measures. More research is needed in this aspect of DEJI Systems Model. To spark interest along this line, some mathematical representations are presented here. One example using plane geometry and projection integrals is presented earlier in Chapter 2. Consider the cliché

Three-dimensional coordinate systems for integration

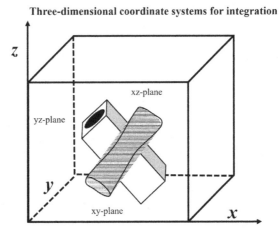

FIGURE 6.1   Geometric dimensions of integration.

of "putting a round peg in a square hole." Where is the alignment with respect to tolerance between the hole and the peg? This could involve a mathematical intrigue of measuring gaps (distance) between the peg and the hole. Conceptually and visually, we can see any peg-and-hole misalignment. But, can we come up with some sort of a mathematical representation, which would be instrumental in comparing pairs of hole and peg, particularly in cases where there are repeated trials. Figure 6.1 presents a geometrical conceptual framework for systems integration. The conventional formula for the distance between two points in a geometric plane can be extended to the following three-dimensional formula for the purpose of measuring the level of mathematical integration:

$$|P_1P_2| = \sqrt{\left(x_2 - x_1\right)^2 + \left(y_2 - y_1\right)^2 + \left(z_2 - z_1\right)^2},$$

where $|P_1P_2|$ is the distance between the points $P_1(x_1, y_1, z_1)$ and $P_2(x_2, y_2, z_2)$.

PLANE GEOMETRY OF SYSTEMS INTEGRATION

We must integrate all the elements of a system on the basis of alignment of functional goals. Systems overlap for integration purposes can conceptually be represented as projection integrals by considering areas bounded by the common elements of subsystems:

$$A = \iint\limits_{A_y A_x} z\left(x, y\right) dy\, dx$$

$$B = \iint\limits_{B_y B_x} z\left(x, y\right) dy\, dx$$

In Figure 6.2, the projection of a flat plane onto the first quadrant is represented as area A while Figure 6.3 shows the projection on an inclined plane as area B. The net projection encompassing the overlap of A and B is represented as area C, shown in Figure 6.4, and computed as

$$C = \iint\limits_{C_y C_x} z(x,y)\,dy\,dx$$

Notice how each successful net projection area decreases with an increase in the angle of inclination of the project plane. The fact is that in actual project execution, it will be impractical or impossible to model subsystem scenarios as

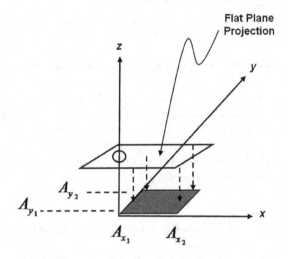

FIGURE 6.2 Flat plane projection for systems integration.

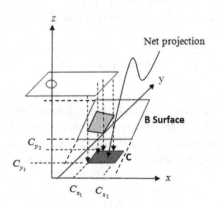

FIGURE 6.3 Inclined plane projection for subsystem alignment and integration.

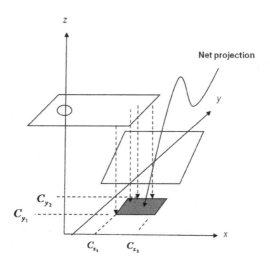

FIGURE 6.4   Reduced footprint of net projection area due to steep incline.

double integrals. But the concept, nonetheless, demonstrates the need to consider where and how project elements overlap for a proper assessment of integration. For mechanical and electrical systems, one can very well develop mathematical representation of systems overlap and integration boundaries.

For the purpose of mathematical exposition, double integrals arise in several technical applications. Some examples are

Calculation of volumes
Calculation of the surface area of a two-dimensional surface (e.g., a plane surface)
Calculation of a force acting on a two-dimensional surface
Calculation of the average of a function
Calculation of the mass or moment of inertia of a body
Consider the surface area given by the integral

$$A(x) = \int_{c}^{d} f(x,y)dy$$

The variable of integration is y, and x is considered a constant. The cross-sectional area depends on x. Thus, the area is a function of x. That is, $A(x)$. The volume of the slice between x and $x + dx$ is $A(x)dx$. The total volume is the sum of the volumes of all the slices between $x = a$ and $x = b$. That is,

$$V = \int_{a}^{b} A(x)dx$$

If substituted for $A(x)$, we obtain

$$V = \int_a^b \left[\int_c^d f(x,y)dy \right] dx = \int_a^b \int_c^d f(x,y)dy\,dx$$

This is an example of an *iterated* integral. One integrates with respect to y first, then x. The integrals with respect to y and x are called the inner and outer integrals, respectively. Alternatively, one can make slices that are parallel to the x-axis. In this case, the volume is given by

$$V = \int_c^d \left[\int_a^b f(x,y)dx \right] dy = \int_c^d \int_a^b f(x,y)dx\,dy$$

The inner integral corresponds to the cross-sectional area of a slice between y and $y + dy$. The quantities $f(x, y)\,dy\,dx$ and $f(x, y)\,dx\,dy$ represent the value of the double integral in the infinitesimally small rectangle between x and $x + dx$ and y and $y + dy$. The length and width of the rectangle are dx and dy, respectively. Hence $dy\,dx$ (or $dx\,dy$) is the area of the rectangle. Thus, the change in area is $dA = dy\,dx$ or $dA = dx\,dy$. For a further conceptual illustration of alignment and integration across platforms, Figure 6.5 shows a series of vertical planes with dots of integration points. The more points are concentrated in a region, the higher the confidence that integration is happening.

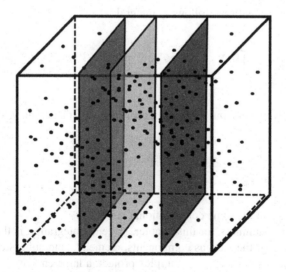

FIGURE 6.5 Dotted pattern of subsystem integration.

COMPUTATIONAL EXAMPLE OF QUANTITATIVE INTEGRATION

Consider the double integral

$$V = \iint_R \left(x^2 + xy^3 \right) dA$$

where R is the rectangle $0 \le x \le 1$, $1 \le y \le 2$. Suppose we integrate with respect to y first. Then

$$V = \int_0^1 \int_1^2 \left(x^2 + xy^3 \right) dy\,dx$$

The inner integral is

$$V = \int_1^2 \left(x^2 + xy^3 \right) dy = \left[x^2 y + x\frac{y^4}{4} \right]_{y=1}^{y=2}$$

Note that we treat x as a constant as we integrate with respect to y. The integral is equal to

$$x^2 \left(2 \right) + x\left(\frac{2^4}{4} \right) - x^2 - \frac{x}{4} = x^2 + \left(\frac{15}{4} \right) x$$

We are now left with the following integral:

$$\int_0^1 \left(x^2 + \frac{15}{4} x \right) dx = \left(\frac{x^3}{3} + \frac{15}{8} x^2 \right)_{x=0}^{x=1} = \frac{1}{3} + \frac{15}{8} = 2.2083$$

Alternatively, we can integrate with respect to x first and then y. We have

$$V = \int_1^2 \int_0^1 \left(x^2 + xy^3 \right) dx\,dy$$

which should yield the same computational result.

In terms of a summary for this chapter, systems integration is the synergistic linking together of the various components, elements, and subsystems of a system, where the system may be a complex project, a large endeavor, or an expansive organization. Activities that are resident within the system must be managed both from the technical and managerial standpoints. Any weak link in the system, no matter how small, can be the reason that the overall system fails. In this regard, every component of a project is a critical element that must be nurtured and controlled. Embracing the systems principles for project management will increase the likelihood of success of projects.

POLAR PLOTS FOR SYSTEMS INTEGRATION

In practical terms, it is important to assess to what extent each subsystem of a system satisfies the desired attributes toward the desired goals of an organization. There are many techniques for conducting such an assessment. This section illustrates the technique of polar plots, which permits a good visual rendition of the comparative analysis. Normally, polar plots provide a means of visually comparing investment alternatives (Badiru, 1991, 2012a, 2012b). However, it can be applied to other organizational pursuits, such as investments in subsystem selection to enhance overall systems integration. In a conventional polar plot, as shown in Figure 6.6, the vectors drawn from the center of the circle are on individual scales based on the outcome ranges for each attribute. For example, the vector for net present value (NPV) is on a scale of $0 to $500,000 while the scale for Quality is from 0 to 10. It should be noted that the overall priority weights for the alternatives are not proportional to the areas of their respective polyhedrons.

A modification of the basic polar plot is presented in this section. The modification involves a procedure that normalizes the areas of the polyhedrons with respect to the total area of the base circle. With this modification, the normalized areas of the polyhedrons are proportional to the respective priority weights of the alternatives, so, the alternatives can be ranked on the basis of the areas of

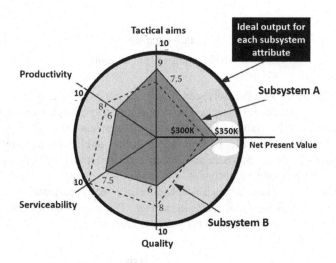

FIGURE 6.6 Basic polar plot.

the polyhedrons. The steps involved in the modified approach are presented
below:

1. Let n be the number of attributes involved in the comparison of alternatives,
   such that $n \geq 4$. Number the attributes in a preferred order $(1, 2, 3,\ldots, n)$.
2. If the attributes are considered to be equally important (i.e., equally
   weighted), compute the sector angle associated with each attribute as

$$\theta = \frac{360^\circ}{n}.$$

3. Draw a circle with a large enough radius. A radius of 2 inches is usually
   adequate.
4. Convert the outcome range for each attribute to a standardized scale of 0 to
   10 using appropriate transformation relationships.
5. For Attribute 1, draw a vertical vector up from the center of the circle to the
   edge of the circle.
6. Measure θ clockwise and draw a vector for Attribute 2. Repeat this step for
   all attributes in the numbered order.
7. For each alternative, mark its standardized relative outcome with respect to
   each attribute along the attribute's vector. If a 2-inch radius is used for the
   base circle, then we have the following linear transformation relationship:
   a. inches = rating score of 0.0
   b. inches = rating score of 10.0
8. Connect the points marked for each alternative to form a polyhedron.
   Repeat this step for all alternatives.
9. Compute the area of the base circle as follows:

$$\Omega = \pi r^2$$
$$= 4\pi \text{ squared inches}$$
$$= 100\pi \text{ squared rating units}$$

Compute the area of the polyhedron corresponding to each alternative. This
can be done by partitioning each polyhedron into a set of triangles and then calculating the areas of the triangles. To calculate the area of each triangle, note that we
know the lengths of two sides of the triangle and the angle subtended by the two
sides. With these three known values, the area of each triangle can be calculated
through basic trigonometric formulas.

For example, the area of each polyhedron may be represented as λ_1 ($I = 1$,
$2,\ldots,m$), where m is the number of alternatives. The area of each triangle in the
polyhedron for a given alternative is then calculated as

$$\Delta_t = \frac{1}{2}\left(L_j\right)\left(L_{j+1}\right)\left(\sin\theta\right),$$

where

L_j = standardized rating with respect to attribute j
L_{j+1} = standardized rating with respect to attribute $j + 1$
L_j and L_{j+1} are the two sides that subtend θ.

Since $n \geq 4$, θ will be between 0 and 90 degrees, and $\sin(\theta)$ will be strictly increasing over that interval.

The area of the polyhedron for alternative i is then calculated as

$$\lambda_i = \sum_{t(i)=1}^{n} \Delta_{t(i)}.$$

Note that θ is constant for a given number of attributes and the area of the polyhedron will be a function of the adjacent ratings (L_j and L_{j+1}) only.

Compute the standardized area corresponding to each alternative as

$$w_i = \frac{\lambda_i}{\Omega}(100\%).$$

Rank the alternatives in decreasing order of λ_i. Select the highest ranked alternative as the preferred alternative.

For readers interested in this comparative technique, Badiru and Lamont (2022) present a comprehensive computational example for illustrative purposes. The problem presented there is used to illustrate how modified polar plots can be used to compare innovation investment alternatives. In that case, the subsystem attributes of interest are quality, profit, productivity, flexibility, and customer satisfaction. It should be noted that the weighted areas for the alternatives are sensitive to the order in which the attributes are drawn in the polar plot. Thus, a preferred order of the attributes must be defined prior to starting the analysis. The preferred order may be based on the desired sequence in which alternatives must satisfy management goals. For example, it may be desirable to attend to product quality issues before addressing throughput issues. The surface area of the base circle may be interpreted as a measure of the global organizational goal with respect to such performance indicators as available capital, market share, capacity utilization, and so on. Thus, the weighted area of the polyhedron associated with an alternative may be viewed as the degree to which that alternative satisfies organizational goals. Some of the attributes involved in a selection problem might constitute a combination of quantitative and/or qualitative factors or a combination of objective and/or subjective considerations. The prioritizing of the factors and considerations is typically based on the experience, intuition, and subjective preferences of the decision-maker. Goal programming is another technique that can be used to evaluate multiple objectives or criteria in systems integration decision problems.

TECHNOLOGY TRANSFER FOR SYSTEMS INTEGRATION

The conventional thoughts, processes, tools, techniques, and strategies of technology transfer are directly applicable to the concept of integration, as advocated by DEJI Systems Model. Why reinvent the wheel for a new system when it can be transferred and adopted from existing wheeled system? The concepts of project management and conversion are useful in planning for the adoption and implementation of a new system. Due to its many interfaces, the area of system adoption and implementation is a good template for utilizing DEJI Systems Model. System managers, engineers, and analysts should take advantage of the effectiveness of project management tools. This applies the various project management techniques that are widely available in the literature. Project management approach is presented within the context of system adoption and implementation for industry development. The Triple C model of communication, cooperation, and coordination is applied as an effective tool for ensuring the acceptance and integration of new technology. The importance of new technologies in improving system quality and operational productivity is discussed.

CHARACTERISTICS OF SYSTEMS TRANSFER

To transfer a system, we must know what constitutes the system. Since systems are often composed of technological assets, some of the discussions are presented in the context of technology transfer. A working definition of technology will enable us to determine how best to transfer it. A basic question that should be asked is:

What is the technology under consideration?
Technology can be defined as follows:
Technology is a combination of physical and nonphysical processes that make use of the latest available knowledge to achieve business, service, or production goals. Essentially, an interplay of people, technology, and processes works to produce products, services, or results. Technology is a specialized body of knowledge that can be applied to achieve a mission or purpose. The knowledge concerned could be in the form of methods, processes, techniques, tools, machines, materials, and procedures. Technology design, development, and effective utilization are driven by effective utilization of human resources and effective management systems. Technological progress is the result obtained when the provision of technology is used in an effective and efficient manner to improve productivity, reduce waste, improve human satisfaction, and raise the quality of life.

Technology all by itself is useless. However, when the right technology is put to the right use, with effective supporting management system, it can be very effective in achieving industrialization goals. That is the essence of integration, as presented by DEJI Systems Model. Technology implementation starts with an idea and ends with a productive process. Technological progress is said to have occurred when the outputs of technology, in the form of information, instrument, or knowledge that is used productively and effectively, lead to a lowering of costs of production, better product quality, higher levels of output (from the same amount of inputs), and higher market share. The information and knowledge involved in technological progress include those which improve the performance of management, labor, and the total resources expended for a given activity.

From a national perspective, particularly from a global perspective, technological progress plays a vital role in improving overall national productivity. Experience in developed countries such as in the United States shows that in the period 1870–1957, 90% of the rise in real output per man-hour can be attributed to technological progress. It is conceivable that a higher proportion of increases in per capita income is accounted for by technological change. Changes occur through improvements in the efficiency in the use of existing technology—that is, through learning and through the adaptation of other technologies, some of which may involve different collections of technological equipment. The challenge to developing countries is how to develop the infrastructure that promotes, uses, adopts, and advances technological knowledge. This is why a systems view of the world, as espoused by this book, is essential at the point in our present digital era.

Most of the developing nations today face serious challenges arising not only from the worldwide imbalance of dwindling revenue from industrial products and oil but also from major changes in a world economy that is characterized by competition, imports, and exports of, not only oil but also of basic technology, weapon systems, and electronics. The emergence of COVID-19 pandemic has further compounded the challenges of global technology interfaces. If technology utilization is not given the right attention in all sectors of the national economy, the much-desired industrial development cannot occur or cannot be sustained. The ability of a nation to compete in the world market will, consequently, be stymied.

The important characteristics or attributes of a new technology may include productivity improvement, improved quality, cost savings, flexibility, reliability, and safety. An integrated evaluation must be performed to ensure that a proposed technology is designed, evaluated, and justified to be integrated into the existing infrastructure, both economically and technically. The scope and goals of the proposed technology must be established right from the beginning of the project. The system "ilities" discussed in the preceding chapter are also applicable for the purpose of technology transfer. An assessment of a technology transfer opportunity will entail a comparison of departmental objectives with overall organizational goals in the following areas:

1. Industrial Marketing Strategy: This should identify the customers of the proposed technology. It should also address items such as market cost of proposed product, assessment of competition, and market share. Import and export considerations should be a key component of the marketing strategy.

2. Industry Growth and Long-range Expectations: This should address short-range expectations, long-range expectations, future competitiveness, future capability, and prevailing size and strength of the industry that will use the proposed technology.

3. National Benefit: Any prospective technology must be evaluated in terms of direct and indirect benefits to be generated by the technology. These may include product price versus value, increase in international trade, improved standard of living, cleaner environment, safer workplace, and higher productivity.

4. Economic Feasibility: An analysis of how the technology will contribute to profitability should consider past performance of the technology, incremental benefits of the new technology versus conventional technology, and value added by the new technology.

5. Capital Investment: Comprehensive economic analysis should play a significant role in the technology assessment process. This may cover an evaluation of fixed and sunk costs, cost of obsolescence, maintenance requirements, recurring costs, installation cost, space requirement cost, capital substitution options, return on investment, tax implications, cost of capital, and other concurrent projects.

6. Resource Requirements: The utilization of resources (human resources and equipment) in the pre-technology and post-technology phases of industrialization should be assessed. This may be based on material input-output flows, high value of equipment versus productivity improvement, required inputs for the technology, expected output of the technology, and utilization of technical and nontechnical personnel.

7. Technology Stability: The maturity of technology matters. Uncertainty is a reality in technology adoption efforts. Uncertainty will need to be assessed for the initial investment, return on investment, payback period, public reactions, environmental impact, and volatility of the technology.

8. National Productivity Improvement: An analysis of how the technology may contribute to national productivity may be verified by studying industrial throughput, efficiency of production processes, utilization of raw materials, equipment maintenance, absenteeism, learning rate, and design-to-production cycle.

NEW TECHNOLOGY MEANS NEW SYSTEMS INTEGRATION

New industrial and service technologies have been gaining more attention in recent years. This is due to the high rate at which new productivity improvement technologies are being developed. The fast pace of new technologies has created difficult implementation and management problems for many organizations. New technology can be successfully implemented only if it is viewed as a system whose various components must be evaluated within an integrated managerial framework. Such a framework is provided by a project management approach.

A multitude of new technologies has emerged in recent years. It is important to consider the peculiar characteristics of a new technology before establishing adoption and implementation strategies. The justification for the adoption of a new technology is usually a combination of several factors rather than a single characteristic of the technology. The potential of a specific technology to contribute to industrial development goals must be carefully assessed. The technology assessment process should explicitly address the following questions:

What is expected from the new technology?
Where and when will the new technology be used?
How is the new technology similar to or different from existing technologies?
What is the availability of technical personnel to support the new technology?
What administrative support is needed for the new technology?
Who will use the new technology?
How will the new technology be used?
Why is the technology needed?

The development, transfer, adoption, utilization, and management of technology is a problem that is faced in one form or another by businesses, industries, and government establishments. Some of the specific problems in technology transfer and management include the following:

- Controlling technological change
- Integrating technology objectives
- Shortening the technology transfer time
- Identifying a suitable target for technology transfer
- Coordinating the research and implementation interface
- Formal assessment of current and proposed technologies
- Developing accurate performance measures for technology
- Determining the scope or boundary of technology transfer
- Managing the process of entering or exiting a technology
- Understanding the specific capability of a chosen technology
- Estimating the risk and capital requirements of a technology

Integrated managerial efforts should be directed at the solution of the problems stated above. A managerial revolution is needed in order to cope with the ongoing technological revolution. The revolution can be initiated by modernizing the long-standing and obsolete management culture relating to technology transfer. Some of the managerial functions that will need to be addressed when developing a technology transfer strategy include the following:

1. Development of a technology transfer plan.
2. Assessment of technological risk.
3. Assignment/reassignment of personnel to implement the technology transfer.

4. Establishment of a transfer manager and a technology transfer office. In many cases, transfer failures occur because no individual has been given the responsibility to ensure the success of technology transfer.
5. Identification and allocation of the resources required for technology transfer.
6. Setting of guidelines for technology transfer. For example,
   a. Specification of phases (development, testing, transfer, etc.).
   b. Specification of requirements for inter-phase coordination.
   c. Identification of training requirements.
   d. Establishment and implementation of performance measurement.
7. Identify key factors (both qualitative and quantitative) associated with technology transfer and management.
8. Investigate how the factors interact and develop the hierarchy of importance for the factors.
9. Formulate a loop system model that considers the forward and backward chains of actions needed to effectively transfer and manage a given technology.
10. Track the outcome of the technology transfer.

Technological developments in many industries appear in scattered, narrow, and isolated areas within a few selected fields. This makes technology efforts to be rarely coordinated, thereby, hampering the benefits of technology. The optimization of technology utilization is, thus, very difficult. To overcome this problem and establish the basis for effective technology transfer and management, an integrated approach must be followed. An integrated approach will be applicable to technology transfer between any two organizations whether public or private.

Some nations concentrate on the acquisition of bigger, better, and faster technology. But little attention is given to how to manage and coordinate the operations of the technology once it arrives. When technology fails, it is not necessarily because the technology is deficient. Rather, it is often the communication, cooperation, and coordination functions of technology management that are deficient. Technology encompasses factors and attributes beyond mere hardware, software, and "skinware," which refers to people issues affecting the utilization of technology. This may involve social-economic and cultural issues of using certain technologies. Consequently, technology transfer involves more than the physical transfer of hardware and software. Several flaws exist in the common practices of technology transfer and management. These flaws include the following:

- Poor fit: This relates to an inadequate assessment of the need of the organization receiving the technology. The target of the transfer may not have the capability to properly absorb the technology.
- Premature transfer of technology: This is particularly acute for emerging technologies that are prone to frequent developmental changes.
- Lack of focus: In the attempt to get a bigger share of the market or gain early lead in the technological race, organizations frequently force technology in many incompatible directions.

- Intractable implementation problems: Once a new technology is in place, it may be difficult to locate sources of problems that have their roots in the technology transfer phase itself.
- Lack of transfer precedents: Very few precedents are available on the management of brand new technology. Managers are, thus, often unprepared for their new technology management responsibilities.
- Stuck on technology: Unworkable technologies sometimes continue to be recycled needlessly in the attempt to find the "right" usage.
- Lack of foresight: Due to the nonexistence of a technology transfer model, managers may not have a basis against which they can evaluate future expectations.
- Insensitivity to external events: Some external events that may affect the success of technology transfer may include trade barriers, taxes, and political changes.
- Improper allocation of resources: There is usually not enough resources available to allocate to technology alternatives. Thus, a technology transfer priority must be developed.

The following steps provide a specific guideline for pursuing the implementation of manufacturing technology transfer:

1. Find a suitable application.
2. Commit to an appropriate technology.
3. Perform economic justification.
4. Secure management support for the chosen technology.
5. Design the technology implementation to be compatible with existing operations.
6. Formulate project management approach to be used.
7. Prepare the receiving organization for the technology change.
8. Install the technology.
9. Maintain the technology.
10. Periodically review the performance of the technology based on prevailing goals.

TECHNOLOGY TRANSFER AND INTEGRATION MODES

The transfer of technology can be achieved in various forms. Project management provides an effective means of ensuring proper transfer of technology. Three technology transfer modes are presented here to illustrate basic strategies for getting one technological product from one point (technology source) to another point (technology sink). A conceptual integrated model of the interaction between the technology *source* and technology *target* is summarized below:

Technology Transfer Modes

- Transfer of technology concepts
- Technology end products

- Technology hardware
- Technology software
- Technology personnel
- Technology blueprint

Modes of Integrated Technology

- Identical technology
- Totally new technology
- New technology concepts or theories

There should always be a reverse technology transfer path through feedforward and feed-back linkages. Technology can be transferred in one or a combination of the following strategies:

1. Transfer of complete technological products: In this case, a fully developed product is transferred from a source to a target. Very little product development effort is carried out at the receiving point. However, information about the operations of the product is fed back to the source so that necessary product enhancements can be pursued. So, the technology recipient generates product information which facilitates further improvement at the technology source. This is the easiest mode of technology transfer and the most tempting. Developing nations are particularly prone to this type of transfer. Care must be exercised to ensure that this type of technology transfer does not degenerate into "machine transfer." It should be recognized that machines alone do not constitute technology.

2. Transfer of technology procedures and guidelines: In this technology transfer mode, procedures (e.g., blueprints) and guidelines are transferred from a source to a target. The technology blueprints are implemented locally to generate the desired services and products. The use of local raw materials and manpower is encouraged for the local production. Under this mode, the implementation of the transferred technology procedures can generate new operating procedures that can be fed back to enhance the original technology. With this symbiotic arrangement, a loop system is created whereby both the transferring and the receiving organizations derive useful benefits.

3. Transfer of technology concepts, theories, and ideas: This strategy involves the transfer of the basic concepts, theories, and ideas behind a given technology. The transferred elements can then be enhanced, modified, or customized within local constraints to generate new technological products. The local modifications and enhancements have the potential to generate an identical technology, a new related technology, or a new set of technology concepts, theories, and ideas. These derived products may then be transferred back to the original technology source as new technological enhancements.

It is very important to determine the mode in which technology will be transferred for systems integration purposes. There must be a concerted effort by all parties to make the transferred technology work within local infrastructure and constraints. Local innovation, patriotism, dedication, and willingness to adopt technology will be required to make technology transfer successful. It will be difficult for a nation to achieve industrial development through total dependence on transplanted technology. Local adaptation will always be necessary.

TECHNOLOGY CHANGE-OVER STRATEGIES

Any development project will require changing from one form of technology to another. The implementation of a new technology to replace an existing (or a nonexistent) technology can be approached through one of several options. Some options are more suitable than others for certain types of technologies. The most commonly used technology change-over strategies include the following:

Parallel Change-over: In this case, the existing technology and the new technology operate concurrently until there is confidence that the new technology is satisfactory.

Direct Change-over: In this approach, the old technology is removed totally and the new technology takes over. This method is recommended only when there is no existing technology or when both technologies cannot be kept operational due to incompatibility or cost considerations.

Phased Change-over: In this incremental change-over method, modules of the new technology are gradually introduced one at a time using either direct or parallel change-over.

Pilot Change-over: In this case, the new technology is fully implemented on a pilot basis in a selected department within the organization.

POST-IMPLEMENTATION EVALUATION

The challenges of systems integration do not end with technology transfer. The new technology should be evaluated after it has reached a steady-state performance level. This helps to avoid the bias that may be present at the transient stage due to personnel anxiety, lack of experience, or resistance to change. The system should be evaluated for the following aspects:

- Sensitivity to data errors
- Quality and productivity
- Utilization level
- Response time
- Effectiveness

With the increasing shortages of resources, more emphasis should be placed on the sharing of resources. Technology resource sharing can involve physical equipment, facilities, technical information, ideas, and related items. The integration of technologies facilitates the sharing of resources. Technology integration is a major effort in technology adoption and implementation. Technology integration is required for proper product coordination. Integration facilitates the coordination of diverse technical and managerial efforts to enhance organizational functions, reduce cost, improve productivity, and increase the utilization of resources. Technology integration ensures that all performance goals are satisfied with a minimum of expenditure of time and resources. It may require the adjustment of functions to permit sharing of resources, development of new policies to accommodate product integration, or realignment of managerial responsibilities. It can affect both hardware and software components of an organization. Important factors in technology integration include the following:

- Unique characteristics of each component in the integrated technologies
- Relative priorities of each component in the integrated technologies
- How the components complement one another
- Physical and data interfaces between the components
- Internal and external factors that may influence the integrated technologies
- How the performance of the integrated system will be measured

CONCLUSION

Technology transfer is a great avenue to advancing industrial and economic development. This chapter has presented a variety of principles, tools, techniques, and strategies useful for managing technology transfer, toward a successful integration. Of particular emphasis in the chapter is the management aspects of technology transfer. The technical characteristics of the technology of interest are often well understood. What is often lacking is an appreciation of the technology management requirements for achieving a successful technology transfer. This chapter highlights both the qualitative and quantitative approaches for technology and systems integration.

REFERENCES

Badiru, A. B. (1991), *Project Management Tools for Engineering and Management Professionals*, Industrial Engineering & Management Press, Norcross, GA.

Badiru, A. B. (2012a), "Application of the DEJI Model for Aerospace Product Integration," *Journal of Aviation and Aerospace Perspectives (JAAP)*, Vol. 2, No. 2, pp. 20–34, Fall 2012.

Badiru, A. B. (2012b), *Project Management: Systems, Principles, and Applications*, Taylor & Francis Group/CRC Press, Boca Raton, FL.

Badiru, A. B., and Gary Lamont (2022), *Innovation Fundamentals: Quantitative and Qualitative Techniques*, Taylor & Francis Group/CRC Press, Boca Raton, FL.

Badiru, A. B., and Marlin Thomas (2013), "Quantification of the PICK Chart for Process Improvement Decisions," *Journal of Enterprise Transformation*, Vol. 3, No. 1, pp. 1–15.

7 Application Case Studies of DEJI Systems Model®

> If there is a reason for everything, everything will have a reason, which is clearly not the case.
>
> —Adedeji Badiru

CASE 1: MITIGATING RISKS AND PREVENTING FAILURE IN PROCESS IMPROVEMENT INITIATIVES BY MIKE KAMINSKI

"We tried that and it didn't stick." How many times have those words been spoken when people within an organization talk about major changes, process improvements, and re-engineering efforts? A *Wall Street Journal* article by Dr. Satya Chakravorty reported the results of a 5-year research mission that showed 60% of all corporate six-sigma initiatives fail. McKinsey & Co. suggests that 70% of complex large-scale change programs fail. According to the Operational Excellence Society, 70% of all process improvement projects fail.

Dissecting failed improvement efforts has been done many times over and provides many root causes and failure modes. Instead of looking to failures for solutions, let's look at a couple of successful initiatives for lessons learned.

At a manufacturing organization, the leadership team was looking for ways to improve organizational effectiveness. The organization has multiple product lines, and each product line has multiple manufacturing facilities. With so many people in so many locations doing so many different things, it was critical to apply consistent methodology and messaging while displaying enough agility to customize solutions to overcome each unique set of challenges.

> **Design** of the project required engaging stakeholders to align and define goals, establish timelines, and determine how success would be measured. Although all the stakeholders agreed that the current state was not acceptable, designing the goals, agreeing on the timelines, and aligning on "what success looks like" required multiple iterations, many revisions, and rich conversations. Although difficult, gaining stakeholder buy-in on the project design was critical to a successful project.
>
> **Evaluation** of the project took center stage from the very beginning. Process improvement initiatives had come and gone, so coming up with objective criteria that could be measured and not disputed became paramount.

DOI: 10.1201/9781003175797-7

Focusing on implementation in this manufacturing environment, the metrics that would evaluate success included on-time delivery, throughput, capacity, cost per unit, and work-in-process inventory.

Justification of the project was an ongoing effort that required continual communication with stakeholders at all levels of the organization. Naysayers and doubters brought plenty of opinions but very few facts or data. By collecting data, measuring results, and communicating with stakeholders on a weekly cadence, the agreed-upon metrics showed that the process improvement efforts and operational changes were making a positive impact.

Integration of the new approaches, methodologies, and tools into the everyday work environment was the only way the improvements and changes were going to have sustainability. Changes would endure if stakeholders at every level were enabled with conditions for success. The workforce was provided visual method sheets and reference guides to help remind them of best practices and standardized methods. The supervisors were provided with scheduling tools, production trackers, and pocket coaching guides to enable them to make quick, accurate decisions throughout the day and week when the availability of materials, people, or machine capacity may change. The leadership team supported facility layout changes that improved workflow and were provided metrics reports that helped them track performance at each facility and spot trends that may require corrective actions.

Results of this initiative increased on-time delivery to 90%, increased average daily output by almost 500%, doubled capacity, decreased manufacturing costs by nearly 70%, and increased WIP inventory turns by 400%.

CASE 2: APPLYING DEJI SYSTEMS MODEL IN PRODUCT DISTRIBUTION BY MIKE KAMINSKI

In a distribution organization, the leadership team was "out of ideas." They were competing in an environment with razor-thin margins and had tried several initiatives to reduce costs with limited success. The distribution network was spread across the country with unique challenges at each site, so it would be important to customize solutions while deploying best practices, consistent methodologies, and sustainable results.

Design of the project started with gaining alignment with the stakeholders. Many of the stakeholders were owners of previous initiatives, so some were supportive, others were not supportive, and a majority chose to "wait and see." Leveraging the supportive stakeholders was critical to developing the goals, establishing project schedules, identifying points of contact, and measuring success. It would be critical to move the "wait and see" stakeholders over to the "supportive" side of the stakeholder alignment continuum for the project to be successful. For the design to be successful, the timeline, measures, and indicators of success would need stakeholder agreement.

Evaluation of the project was defined to measure benefits, track changes to costs, and understand the impact on people. Implementation costs were also tracked to make sure that investments in improvements had acceptable rates of return. Metrics that would measure success in this environment included cost savings, cultural transformation, workplace organization, and bonus payouts.

Justification was an ongoing effort. The first scope of work was limited to an assessment. The goal of the assessment was to help with stakeholder alignment and quantify the size of the opportunity. Answering questions like "how much money could be saved?," "how could the workforce culture be improved?," "what could be done to organize the workplace better?," and "how would all that make a positive impact on bonus payments?" The assessment more than justified the cost of starting a pilot project and convinced some of the stakeholders that were "on the fence" that process improvement ideas from the assessment would provide organizational benefits. Once the pilot project commenced, justification activities closely monitored results, tracked benefits, and communicated progress. Due to the justification diligence, the pilot project blossomed into four additional projects that encompassed the entire distribution network.

Integration of best practices, lean concepts, 5-S workplace organization, and improved workflow into the culture of the organization was the only way the initiative would be sustainable and not fall victim to the "flavor of the month" that had plagued many previous initiatives. Best practices were integrated by making sure that methodologies became part of new hire training and deploying visual method sheets to remind employees in the workplace of the "way we do it here." Lean concepts were integrated by providing training to every employee in the distribution centers. Training was customized in a way that resonated with each person's area of responsibility and translated for employees that considered English as a second language. Work centers were organized using the 5-S structure, and employees were part of the process. We needed to make sure that changes were happening "with them," not "to them." Workflow changes may have been some of the most disruptive process improvements, so to gain support, and assure sustainability the metrics were tracked closely and communicated frequently. As the workforce, supervisors, and leadership saw tangible improvements, momentum picked up, and sustainment was not a problem.

Results of this initiative achieved annualized cost savings of more than $11 million, transformed the culture to a lean workplace, trained more than 4,000 employees in lean principles, completed 5-S on more than 4 million square feet of workspace, and enabled employees to increase their bonuses by 5%–15%.

Both case studies were very successful implementations that were sustained. In both cases, thoughtful and deliberate efforts went into the design, evaluation, justification, and integration activities that are represented in the DEJI systems framework. Consider adding the DEJI model to your next process improvement implementation plan to help assure that your efforts are part of the 30%–40% of initiatives that succeed.

ABOUT THE AUTHOR

Mike Kaminski leads the Operations Excellence practice at Accenture Federal Services. He is an organizational leader, continuous improvement professional, and lean practitioner with recognized experience leading organizational transformations and workforce optimizations in operations, engineering, and quality disciplines within the manufacturing, defense, distribution, process, and retail industries. He has significant experience with workforce effectiveness, lean implementations, performance management, productivity improvements, engineered labor standards, workplace organization, activity-based staffing, Malcolm Baldrige National Quality Award, high-performance teams, QS 9000 registration, and plant start-ups. Mike has a Bachelor of Science degree in Industrial and Systems Engineering from The Ohio State University and a master's in Business Administration from California Coast University.

CASE 3: BUSINESS CLIMATE AFFECTING POWER SUPPLY: A TOTAL SYSTEMS PERSPECTIVE BY ADEDEJI BADIRU

Consider the unfortunate case of electricity supply failures, which can be due to a variety of reasons, including weather incidents, accidental events, mismanagement issues, technical difficulties, or deliberate acts. We often focus, rightfully so, on the technical aspects impinging upon power-supply systems. While there are noticeable technical impediments that prevent consistent and reliable power supply, there are many other factors of concern, including the business climate considerations. We must look at power supply, holistically, from a total systems perspective. In this respect, a system is a collection of interrelated elements, whose collective output together is higher than the sum of the individual outputs of the elements. Thus, the business climate should be a part of the considerations for resolving power-supply problems.

There are attributes, factors, and indicators for assessing a supportive business climate. If an assessment points to a laxity in the business climate, then we can take corrective actions to remedy the adverse findings. We can put the best technical system in operation, but if the business climate entails an arrant display of inefficiency, the overall technical system can fail.

TEXAS CASE EXAMPLE

The February 2021 power failure in the state of Texas (United States), brought on by an unexpected (and unaccustomed) cold weather, gives credence to the need to pay more attention to the business climate. While the technical inability of the power plant to function properly in an extreme cold weather is the obvious attribute to point fingers at, the problem should not have been a total surprise because there were indicators, all along, in the power-supply business climate of Texas. Namely, the closed-loop unregulated business of supplying energy to Texas. An unregulated business climate could create a free-form shenanigan practice that does not safeguard against adverse weather occurrences. More can be researched and discussed about this case example as the basis for a template of best practices for the power industry.

OHIO CASE EXAMPLE

A widely reported 2020 case of chicanery in the nuclear power industry in Ohio is of relevance to the caution about business climate consideration for the power supply industry. The case involved an alleged complex corruption-driven high-dollar bribery scandal that reached all the way to the state government, for the purpose of manipulating dollars and cents in the nuclear industry through government subsidies and bailouts. In the final analysis, the end-point consumers suffer either through shackled supply or ridiculously high pricing. Ohio is reported not to be alone in the nuclear energy business climate scandals. Similar cases have been reported in Illinois and other places.

The examples from developed-nation transgressions don't bode well for what can be expected in the looseness of developing countries, unless we tighten our belts and processes upfront, which is, precisely, the purpose and focus of our NAE Power Committee.

A RECOMMENDED SYSTEMS FRAMEWORK

Based on the preceding discussion, it is recommended that we take a look at some of the systems engineering tools and techniques available in the literature (see Badiru, 2019). Of particular importance is systems integration to link the technical elements of power supply to the business elements. One model that is aptly relevant for that purpose is the trademarked **DEJI Systems Model®** that is represented by the figure below. A condensed summary of the essential application of the model to a power-supply scenario is described below:

Design: Design of the technical aspects of power plants (whether renewable or non-renewable)

Evaluation: An assessment of power-supply needs vis-à-vis the power-plant capabilities

Justification: A hybrid review of power-supply flow to areas of need (home, business, and industry)

Integration: Mapping of power-supply realities to human behavior in consumption and conservation

A total systems approach facilitates a consideration of all the nuances of power conception, generation, transmission, distribution, consumption, and conservation.

Also, regulatory, legal, international standards, industry consensus, human needs, and contractual expectations can be factored into the total systems thinking framework.

CASE 4: OPERATION AND MANAGEMENT OF AN R&D TECHNOLOGY VILLAGE BY ADEDEJI BADIRU

SYSTEMS INTEGRATION FOR OPERATION AND MANAGEMENT EXCELLENCE OF A PROPOSED R&D TECHNOLOGY VILLAGE

It is through the process of integration that a proposed R&D technology village can be aligned consistently and productively with the normal pattern of work in the local context. If we follow the systems framework presented in DEJI Systems Model, the "integration" stage will convey the sustainability of the village in living up to the operational conceptualization, evaluative metrics, and value justification for the village. In particular, if collaborative operations are expected with academia and industry, certain rubrics must be demonstrated with respect to the systems "ilities" of the village. This write-up presents guidelines for operation and management excellence of the technology village. Thus, answering questions rather than posing new questions. This new write-up is more descriptive and prescriptive rather than inquisitive.

USING ENGINEERING PROBLEM-SOLVING METHODOLOGY

In as much as we are practicing engineers, we should use our typical engineering problem-solving methodology to implement the systems framework advocated for the technology village. This is the best way to sell our proposal to politicians and non-engineers. Unlike many problem-solving approaches, the engineering problem-solving methodology uses the application of mathematics and the sciences to create useful solutions to practical problems.

Engineers translate scientific knowledge and discoveries into practical applications, innovations, and inventions that benefit the society. Engineers don't just predict future events; they use technical strategies to facilitate positive future events. The engineering field of industrial engineering is particularly well versed in this regard for the proposed operation and management of a technology village.

TECHNOLOGY VILLAGE OPERATION VIA INDUSTRIAL ENGINEERING

An industrial engineer is one who is concerned with the design, installation, and improvement of integrated systems of people, materials, information, equipment, and energy by drawing upon specialized knowledge and skills in the mathematical, physical, and social sciences, together with the principles and methods of engineering analysis and design to specify, predict, and evaluate the results to be obtained from such systems. This is exactly what a technology village needs in incorporating the multi-faceted requirements for operating and managing a technology village. The typical eight-step process for engineering problem solving is presented below. Depending on the specific problem at hand, the steps can be condensed or expanded to cover all the requirements:

Step 1: Materials needed
Step 2: Explicit problem statement
Step 3: Identification of what is known and unknown
Step 4: Specification of assumptions and circumstances
Step 5: Schematic representations and drawings of inputs and outputs
Step 6: Engineering analysis using equations and models
Step 7: Articulation of the results in a presentable format
Step 8: Verification, presentation, and "selling" of result

The good thing about the engineering process is that technical, social, political, economic, and managerial considerations can be factored into the process. The end justifies the details at hand. Based on the recommended approach of tackling problems from a systems perspective, we add an integration requirement to the engineering problem-solving steps. Integrate the engineering operation and management into the normal operating procedures of the technology village. It is through systems integration that a sustainable actualization of the result can be achieved as a contribution to national development. Thus, the answers and strategies for the systems "ilities" and "lities" relevant to the village are addressed in a tabulated format in Table 7.1.

The foregoing narratives can be combined into the visual schematic of the DEJI systems model® provided in Figure 7.1. The bulleted elements under the metrics touch on only a starting set of desirable metrics. Coalescing the inputs and thoughts of the technology village committee members may lead to additional relevant metrics to include. Essentially, the framework can be expanded as comprehensive as desired to cover all the important operation and management aspects of the technology village. The advantage of using the framework provided is that all the pertinent issues can be compactly included so that most things are accounted for. Further, future expansion of the framework will allow additional "future" considerations to the included as they develop.

CASE 5: IMPROVING CYBERSECURITY BY USING SEMI-QUANTITATIVE ANALYSIS WITHIN THE DESIGN, EVALUATION, JUSTIFICATION, AND INTEGRATION (DEJI) MODEL BY JOHN S. BOMMER, JR., J.D., CISM, LT. COL., USAF (RET.); SAE G-32 VOTING MEMBER, CYBERSECURITY, CYBER-SUPPLY CHAIN RISK MANAGEMENT, AND CYBER RISK ASSESSMENTS SME

INTRODUCTION

Cyberspace is the domain in which people, nation-states, private sector companies, and the Department of Defense (DoD) attempt to apply security applications,

TABLE 7.1
Alignment and Integration of R&D Technology Village

Category of "ilities"	Requirements	How to Accomplish
Adaptability	The village needs to embrace the needs of industry and academia.	Offer in-service internships and work exchanges for university students and junior engineers from industry.
Affordability	Keep the village operation within cost benchmarks of similar tech centers.	Create revenue-generating technical service partnerships with universities and local industry.
Agility	Keep village equipment current and maintained to adapt to new technological developments.	Source for equipment donation from high-tech industries and government centers.
Dependability	If the village is dependable, it will be trusted as a think tank for technology developments.	Create village-university-industry-government alliances to anchor the technology village to R&D platforms throughout the nation.
Desirability	Keep the village relevant for national areas of need.	Monitor national strategic plans and align products of the village to the nation's desirable outputs.
Maintainability	Institute a maintenance culture to keep the village vibrant and responsive.	Establish periodic review, repair, and refresh on a regular schedule.
Modularity	Make the village switch components in response to new technology realities.	Don't over-fix equipment installations that may limit interchangeability of R&D segments.
Operatability	Ensure that the village will be able to continue operation in accordance with its stated mission.	Provide adequate power, water, and Internet resources to facilitate consistent operation.
Practicality	Align products of the village to service and results that serve the needs of the local population.	Invite local entrepreneurs and small-scale industry operators to participate in practical product development competitions.
Quality	Embrace and practice industry standards for quality of services and products.	Use industry tools of Six-Sigma and Lean Principles.
Reachability	A good technology center that is not reachable is of no use to the envisioned technological advancement of the nation.	Provide world-class transportation access roads and facilities leading to the tech village.
Reconfigurability	Make the village operations flexible to take on new R&D needs.	In times of national crisis, make the village respond to current needs. For example, quickly doing R&D to respond to a pandemic or national security issues.

(Continued)

TABLE 7.1 (CONTINUED)

Reliability	The best R&D organizations are those that have reliable operations.	Ensure that the management infrastructure of the village demonstrates consistent go-to reliability in contributing to the technological advancement of the nation.
Testability	Produce products that can be technologically affirmed through test and evaluation techniques to match industry's testing standards.	Engage in reverse engineering to advance R&D activities of the village.
Transmittability	For technological relevance, make digital products of the village available as "creditable" open source assets for R&D.	Digital data, products, and other intellectual outputs of the village need to be made available for achieving credibility in the R&D communities.

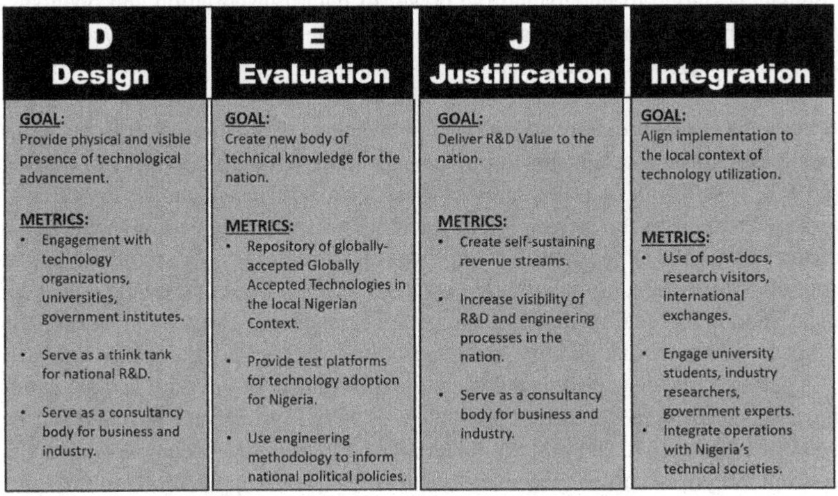

D **Design**	**E** **Evaluation**	**J** **Justification**	**I** **Integration**
GOAL: Provide physical and visible presence of technological advancement.	**GOAL:** Create new body of technical knowledge for the nation.	**GOAL:** Deliver R&D Value to the nation.	**GOAL:** Align implementation to the local context of technology utilization.
METRICS: • Engagement with technology organizations, universities, government institutes. • Serve as a think tank for national R&D. • Serve as a consultancy body for business and industry.	**METRICS:** • Repository of globally-accepted Globally Accepted Technologies in the local Nigerian Context. • Provide test platforms for technology adoption for Nigeria. • Use engineering methodology to inform national political policies.	**METRICS:** • Create self-sustaining revenue streams. • Increase visibility of R&D and engineering processes in the nation. • Serve as a consultancy body for business and industry.	**METRICS:** • Use of post-docs, research visitors, international exchanges. • Engage university students, industry researchers, government experts. • Integrate operations with Nigeria's technical societies.

FIGURE 7.1   DEJI Systems Model applied to R&D technology village.

techniques, coding, and procedures to gain a measure of privacy with the data that they own or have created. This attempt at providing a measure of privacy is combining the cyberspace domain with security. Thus, our global society has created the term "cybersecurity."

It has been nearly 40 years since the word "cyberspace" first appeared in print, in a short story by William Gibson for the July 1982 edition of the now-defunct science-fiction magazine, Omni. In an interview in the Paris Review, Gibson describes how he came up with the word "cyberspace." He stated:

> the first thing I did was to sit down with a yellow pad and a Sharpie and start scribbling—infospace, dataspace. I think I got cyberspace on the third try, and I thought, oh, that's really weird word. I liked the way it felt in the mouth—I thought it sounded like it meant something while still being essentially hollow.[1]

From Gibson's own words, cyberspace was nothing more than a coined thought from his individual mind. His thought of cyberspace was to see it as a virtual reality realm.[2]

Within cyberspace concept being formed from individual thought, the U.S. Military and DoD (and eventually the world) took this idea and super-imposed the concept onto an open systems inter-connection model that was created by the Defense Advanced Research Projects Agency (DARPA). Cyberspace as DARPA created it was never intended to have privacy. The telecommunications and computer communication medium was intended to share information and allow research collaboration. It was not until the U.S. Government and private sector companies started to understand the power of data transportation and open systems collaboration through this new medium did the concepts of privacy and security begin to raise their head.

Because of the recent attacks like Solar Winds[3] and Colonial Pipeline,[4] there needs to be a standard premise put forth of "what is cybersecurity?" Is it the application of security mechanisms to preserve the integrity and availability of a cyber physical system? Or, is it the application of security mechanisms in a systematic process to improve the privacy of the data that moves through components? This chapter posits that cybersecurity is improved by increasing privacy, confidentiality, and safety by using semi-quantitative analysis (SMQA) within the quality enhancement model for quality that has an iterative, recursive assessment during the design, evaluation, justification, and integration (DEJI) of product development.[5]

Therefore, cybersecurity is not based on the concept of integrity and availability. Cybersecurity is based on the concept of "Privacy." Privacy is based on the concept of an individual's right to be let alone.[6] Thus, when society discusses the concept of cybersecurity, cybersecurity attributes are the application security protocols, controls, tactics, techniques, processes, procedures, and coding to gain a measure of privacy with the data and computer components, telecommunications, and satellite communications that an individual, company, state, or nation-state has created. In sum, SMQA within a cybersecurity risk assessment analysis approach using the DEJI quality framework can enhance cybersecurity attributes because these combined approaches and framework can systematically increase software, hardware, firmware, and a cyber physical system's security posture for privacy.

SEMI-QUANTITATIVE ANALYSIS

Microelectronics Trustworthiness is positioned as the call to create better hardware, software, and firmware assurance for Cyber Physical Systems and Industry

4.0 migration. National Defense Authorization Act (NDAA), section 224[7] has produced a mandate for DoD, Federal, and private sector original equipment manufacturers (OEMs) to collaborate and improve the assessment and security application process for threats, weaknesses, and vulnerabilities (TWVs) within a cyber component bill of materials.

To resolve the TWVs, many organizations have been trying to actively resolve the system within systems security problems through quantitative analysis procedures. They have been grappling with developing a systematic, quantitative analysis that will allow for facilitating an assurance posture that tackles supply chain risk management, cyber-supply chain risk management (C-SCRM), and confidentiality, integrity, and availability regarding cyber physical systems (CPS) that create and supply the Automotive, Aerospace, Industry 4.0, and Weapons Systems arenas.

The most widespread means of handling TWVs in hardware, software, firmware, and CPS[8] has been doing a quantitative risk analysis that investigates the weaknesses and vulnerabilities associated with the hardware, software, and firmware of the CPS. The theory being presented here is that components[9] require a SMQA process for enhancing the cybersecurity posture of a CPS within a system of systems.

Thus, the premise of SMQA is that the operational verification and validation of a CPS should be done by combining stakeholder qualitative analysis with a CPS operations-focused, quantitative analysis of the systems hardware, software, and firmware. The analysis process uses transparency, where the mission and profit value assumptions of the CPS are defined by the stakeholder and how those value assumptions are applied is imperative, in order to make the analysis relevant. This can be done in reasoned, qualitative valuation of variables, where the stakeholder defines what is valuable to the mission and profit.[10] After qualitative valuating, framing, scoping, and selecting the target assurance level for the CPS, the quantitative cyber risk assessment analysis should be done based on the mission-defined parameters of the MBSE[11] and test and evaluation plan[12] of the validation requirements. In addition, the CPS should be operated and evaluated as part of the system of systems verification.

While quantitative risk analysis reports have been produced and widely used up to now, there is ambiguity in how these quantitative analysis reports are actually read and used for the health, safety, and security application of decision-makers.[13] Decision-makers frequently do not apply a fully rational approach to digesting the data provided but optimize decision-making within constraints or a bounded rationality.[14] "Decision-makers spend equal time and high-level attention on all parts of a presented quantitative data report, but...the participants' perceptions of understandability and helpfulness" of the reports are related to how the material is presented.[15] It is also documented that presentation methods, such as graphs and tables, may influence the interpretation of quantitative data. For instance, describing a glass as half full may convey an optimistic outlook, while describing it as half empty has a more pessimistic tone.[16] On that same edge, specific types of graphs may be understood better, and, the other way round, the presentation format of research findings may lead to misinterpretations.[17] Thus, framing and scoping the problem set becomes very important function for the unbiased analysis of a system of systems with multiple CPS, because it limits ambiguity and allows for targeted analysis.

However, straight quantitative analysis supporters may disagree that framing and scoping can assist with proper analysis. The effects of framing on decision outcomes have usually been seen as signs of irrationality. According to the so-called principle of invariance (or extensionality), "the preference order between prospects should not depend on the manner in which they are described."[18] Shlomi Sher and Craig R.M. McKenzie pointed out that it is normatively unproblematic for logically equivalent descriptions to yield different decisions if they convey different information to the decision-makers. However, it can reasonably be required that decision-makers should make the same decision in two different frames if these frames "convey exactly the same information about choice-relevant pieces of information."[19] But, this does not always happen because there is a value stream that is part of the decision factors. According to the value management model, value is where the function (natural or characteristic action performed by a product or service) is divided by the cost.[20] Here, value is equal to the reliable CPS performance to meet customer's desired requirements and the goal is to increase return on investment to the stakeholder's enterprise. Hence, a main focus of CPS analysis should be on functional mission analysis and function worth, where the objective is to increase value to customers and stakeholders.

In summary, the crux of the matter is that SMQA is a sound analysis approach for enhancing the ability to mitigate and secure the threats, weaknesses, and vulnerabilities of a system of systems. The SMQA approach on hardware, software, and firmware allows for the creation of a Cyber Physical System Security Plan (CPSSEP) that has the qualitative variable defined by the stakeholders and the quantitative risk scoped under the parameters of mission assurance/profit capability, with safety and cybersecurity as defined constants and constraints. Then, the decision options generated from this combined analysis process will only be evaluated in terms of its consequences and other decision-relevant characteristics set by the stakeholders' pre-quantitative analysis framing, so that the presentation of the material does not lend itself to creating a decision bias. In addition, verification and validation testing should allow the combined SMQA parameters to draw the conclusion needed for the particular environment where the CPS is destined to operate, because argument-based tools can be used either to complement probabilistic risk analysis or to substitute it when the available information is insufficient for formal risk analysis.[21] Thus, the SMQA approach should encompass a qualitative variable that is decided by the stakeholder with a defined scale based on targeted assurance level that prevents mission degradation or hazards. Then, use the quantitative analysis to provide the specified operational requirements testing and system security specification verification/validation testing for the CPS, hardware, software, and firmware.

CYBER RISK ASSESSMENT ANALYSIS APPROACH

The cyber risk assessment (CRA) analysis approach uses a five-step cyber risk assessment analysis process that has been adapted from NIST Publication 800-37 Rev2; ISO/SAE 21434; ISO 26262; and ISO 31000[22]. The five-step cyber risk assessment analysis approach:

1. Identifies critical infrastructure and key resources (CIKR)[23] safety and security requirements of stakeholder
2. Focuses on identification of CIKR assets
3. Identifies the mission boundary and security perimeter by mapping enclave connectivity
4. Identifies the attack surface of the subsystems based on the mission boundary and security perimeters of enclave components
5. Uses a TWV feasibility assessment methodology that applies the safety hazard analysis and risk analysis (SAHARA)

The CRA analysis approach affords the assessment and operations stakeholders the ability to conduct a functional mission analysis (FMA). The FMA is a framework for examining mission assurance and uses a top-down approach that focuses on identifying and understanding cyber operations interactions and communications behaviors that can lead to loss.[24] This analysis approach promotes an asset vulnerability feasibility study to analyze hardware, software, cyber -physical systems, cross-sector horizontal and vertical logistics integration and interdependencies to include cyber-related issues[25] that reveal themselves based on the FMA. One of the key attributes in using FMA concepts is to start with stakeholder and mission research to understand the safety and security concerns. Then, the analysis approach requires an architectural analysis with the stakeholders identified subject matter experts (SMEs) and points of contact (POCs). Architectural analysis is the activity of discovering important system properties using conceptual and physical models of the systems of interest.[26] After the architectural analysis, the CRA involves an iterative investigation where interviews and question/answer sessions are done with the stakeholder's SMEs/POCs, because each system architecture is going to be different and will have multiple hardware, software, and cyber-connectivity points for every enclave and mission. Thus, this architectural analysis step of the cyber risk assessment analysis approach focuses on how to dissect, understand, and propose allocation of resources to improve the security posture connections between enclave designs within the mission boundary.

Once the architecture is deciphered, the mission boundary and security perimeter are determined to better secure the connection between enclave designs and promote secured system operations within the CIKR environment, as illustrated in Figure 7.2.

A "security perimeter" refers to natural barriers or built fortifications to either keep intruders out or to keep captives contained within the area the boundary

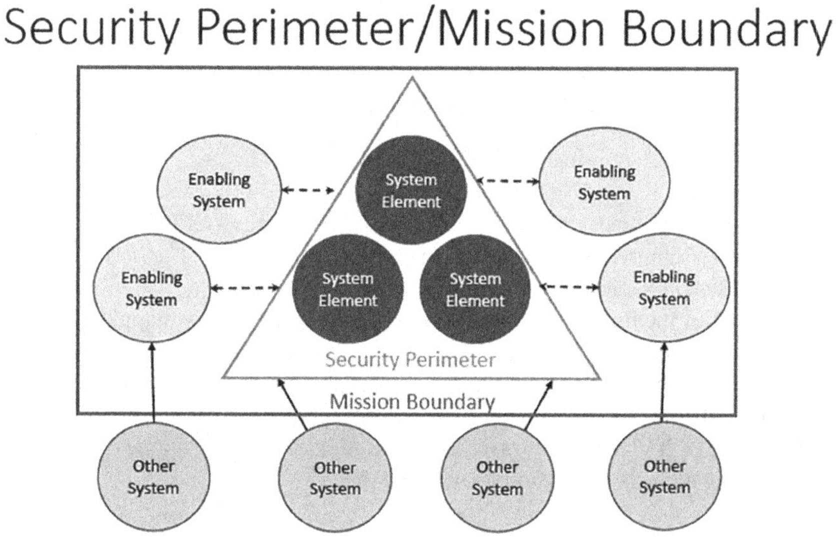

Security Perimeter/Mission Boundary

FIGURE 7.2   Security perimeter and mission boundary.

surrounds.[27] A security perimeter[28] is originally a military strategy that seeks to delay rather than prevent the advance of an attacker by producing protection mechanism consisting of an obstacle or a hindrance to movement on land or over water as an initial protection mechanism against unwanted and undesired actions by a person, party, or nation-state army. Here, the placement of protection mechanisms is intended not only to decrease the movement but also to provide the owner of land, sea, space, or cyberspace time to detect and respond to an attack and so reduce and mitigate the consequences of an unwanted and undesired breach.

The CPS system security perimeter,[29] see Figure 7.2, is where the stakeholder has defined the minimum cybersecurity and safety requirements to have hardware assurance (HwA), software assurance (SwA) of their CPS, software, and firmware[30] during product life cycle production. Meeting this minimum cybersecurity and safety requirement means that system elements and CPS components have the SCRM processes and procedures to provide the provenance[31] and pedigree[32] of the hardware bill of materials (HBOM[33]), software BOM (SBOM), firmware, and CPS BOM's chain of custody. In addition, system elements and enabling systems must meet the required assurance level and provide proof of their security. This proof is provided by using the zero-level trust methodology, where access is granted because security boundary requirements are demonstrated/verified/validated. See Figure 7.3.[34]

The data entry and system connection into the CPS security perimeter by any system element or subsystem asset means that the element and system asset have the required application security.

Thus, the HBOM, SBOM, firmware, and CPS BOM's chain of custody establishes a degree of assurance that the CPS, software, and firmware's content,

The C-SCRM Zero-Trust Model

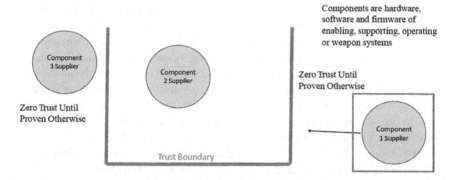

FIGURE 7.3   The C-SCRM Zero-Trust model.

function, and integration operate in the intended manner and are free from being exploited from outside the CPS security perimeter.

Lastly, a collaborative cyber-resilience SAHARA threat feasibility model is used to help define the appropriate cyber security posture to address CIKR stakeholders' needs.[35] The analysis approach relies on the feasibility assessment to prioritize asset allocation and then proceeds with attack path analysis and vulnerability feasibility assessment to do the initial scoping of a cyber risk assessment, in order to ensure safety of CIKR operations.[36]

In a nutshell, the five-step cyber risk analysis process affords an opportunity for the stakeholder and assessor to comprehensively digest and decide TWVs that can be avoided through an SMQA approach that properly allocates assets, so systems and their components can have cybersecurity systematically added in the life cycle and iteratively applied during the system engineering process using the DEJI model.

DEJI MODEL USE OF SMQA IN CYBER RISK ASSESSMENT

In the DoD system engineering process, there is an expressed desire by the weapon systems acquisition program management offices to have an iterative review process for adding security and cybersecurity to hardware, software, and CPS development in production. The DEJI model is unique among product development tools and techniques because it explicitly calls for a re-justification of the product within the product development life cycle and the DEJI model facilitates such a recursive design-evaluate-justify-integrate process for product evolution feedback looping.[37] An iterative and recursive process is important in SMQA for cybersecurity because components and cyber physical systems builds are under an ever-changing, metamorphosis. System designs and functionality are dependent on mission needs and stakeholder requirements. Those needs and requirements are met through integration of software, hardware, and systems that are built from multiple physical system original equipment manufacturers and software source-code data sources. This integration of multiple physical components and data sources can cause horizontal and vertical integration issues throughout the life cycle of a CPS, weapon system, and CIKR, because threats, weaknesses, and vulnerabilities that cause degradation reveal themselves during integration. SMQA is designed to combine stakeholder qualified desires with data quantified requirements in the cybersecurity risk assessment process, which can then be applied within the DEJI product integration recursively, until justification for each mitigation used to address TWVs that hamper and impede the intended safety and security of operations are tested and validated. These combined methodologies create the ability to achieve resiliency because they are applied during research and development and the creation of product. Presented below are some questions that can be used to trigger the combined use of the above frameworks.

- How can the cyber physical system be attacked and what are the impacts of the attacks?
- What are the threats to safety, if the system functionality is degraded?
- What is the cybersecurity posture needed for mission assurance?
- How can the threats come through the information exchange points of the cyber physical systems?
- How can threats be deterred/mitigated from impacting the safe operations of cyber physical systems or manufacturing systems?
- How can cybersecurity be implemented in the manufacturing process of system components?
- Is there policy in place for responsibility and accountability for cybersecurity in the product integration plan?
- Is there visibility on how cybersecurity is applied in development and production of the cyber physical system supporting the IoT, IoS, or IoD?

By asking these kinds of questions during the systems engineering process, the frameworks provide an ability to extrapolate, to document, and to mitigate cybersecurity and safety concerns in supply chain risk management (SCRM) and in cyber-supply chain risk management (C-SCRM) for research and development, operations technology, and maintenance phases of the Industry 4.0 CPS or weapons system life cycle process. See Figure 7.4.[38]

CASE 5 SUMMARY

This chapter is intended to illuminate a new iterative approach to securing hardware, software, CPS, and weapon systems. Use of the DEJI Model with SMQA can improve cybersecurity by enhancing the five-step cyber risk assessment analysis process ability to recursively address SCRM and C-SCRM within product development. The key to success for SCRM and C-SCRM securing HW, CPS, and WSP is the integration of cybersecurity attributes where those attributes do not degrade the mission and iteratively testing all elements of the systems to ensure

Cybersecurity & Safety

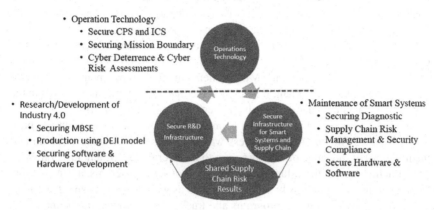

- Operation Technology
  - Secure CPS and ICS
  - Securing Mission Boundary
  - Cyber Deterrence & Cyber Risk Assessments

- Research/Development of Industry 4.0
  - Securing MBSE
  - Production using DEJI model
  - Securing Software & Hardware Development

- Maintenance of Smart Systems
  - Securing Diagnostic
  - Supply Chain Risk Management & Security Compliance
  - Secure Hardware & Software

FIGURE 7.4   Cybersecurity and safety.

that systems function and operate to meet mission and stakeholder requirements while maintaining and sustaining the desire security posture and assurance level.

CASE 6: DEJI SYSTEMS MODEL IN CURRICULUM MANAGEMENT

From a systems implementations standpoint, the DEJI model was successfully applied to the following two high-level strategic decisions at the Air Force Institute of Technology (AFIT):

1. Program consolidation at the department level
2. Establishment of a Thesis Processing Center

In the first example, the systems-based approach to program consolidation was successfully used for curriculum integration by the Department Head of the Department of Systems Engineering & Management in 2011. Figure 7.5 illustrates how the elements of the integrated curriculum were factored into an enhanced curriculum that caters to the various student stakeholders of the department and mitigates the previously siloed delivery of options in the department. The nomenclature in the figure are ENV (Department Code), GES (Graduate Environmental & Science Engineering program), GEM (Graduate Engineering Management program), GSE (Graduate Systems Engineering program), CE GEM (Civil Engineer GEM program), GRD (Graduate Research & Development program), GIR/ESI (Graduate Information Resources program with Enterprise Integration Track), GCA (Graduate Cost Analysis program), GSE DL (Distance Learning delivery of GSE program), OpTech (Certificate program for Operational Technology), and HSI Track for 711th Human Performance Wing (Specialized Track for Air Force).

The HSI track is a special-delivery program that caters to the specific workforce development requirements for an external customer by leveraging an integrated coalescing of program elements that exist in the department. In the process of doing a systems-based remodeling of the programs, it was found necessary to integrate, eliminate, or redesign some of the legacy programs. In fact, as a part of the systems-informed process, the department's name was changed from the Department of Systems and Engineering Management to the Department of Systems Engineering and Management. It is remarkable that a simple tweak of words in the department's name could help resolve issues of stakeholder concerns.

Without an over-arching systems approach, the departmental would not have been possible. The key to this process was an in-depth analysis (evaluation and justification) of the value of each program to the overall department mission. With a clear delineation of the values (or lack thereof) of some programs, it was possible to justify changing previously sacred programs as well as realize synergistic relationships among the various programs and tracks. The following are some of the guidelines and important questions facilitated by using the DEJI model approach to curriculum integration: What are the unique characteristics of each component in the integrated curriculum? How do the characteristics complement one another? What physical interfaces exist among the components? What data and information interfaces exist among the components? What ideological differences exist among the components? What are the data flow requirements for the components? What internal and external factors are expected to influence the integrated curriculum? What are the relative priorities assigned to each component of the integrated curriculum? What are the strengths and weaknesses of the integrated curriculum? What resources are needed to keep the integrated curriculum operating satisfactorily? Which organizational unit has primary responsibility for the integrated curriculum? The DEJI model encourages better alignment of curriculum design with future academic needs. The stages of the model require research for each new curricular element with respect to DEJI. Existing analytical tools and techniques can be used at each stage of the model, regarding such topics as value-added benefits, justification, and integration of specific components of a curriculum. Figure 7.5 illustrates the pathways to curriculum consolidation and integration.

In the second example, the DEJI model was utilized in the establishment of the Thesis Processing Center (TPC) at AFIT in 2013. Prior to TPC, theses and dissertations were processed in paper formats. TPC converted the old system to an electronic processing system, which facilitates faster and more reliable dissemination of AFIT thesis research products. The classical conventional wisdom was that AFIT did defense-focused research, some of it at the classified level. Thus, there was no push to make the products electronically available for fast and wide distribution. Using the DEJI model approach, the Dean was able to carry the AFIT team through the stages of TPC design, evaluation, justification, and integration. Today, while there is no physical "four-walled center," the functions of TPC are "integrated" into the functions of single-staff member. As thesis processing workload increases, the systems approach allows for a review and redesign of TPC, which is an ongoing iterative process.

DEJI Application to Academic Program Realignment And Consolidation (PRAC)

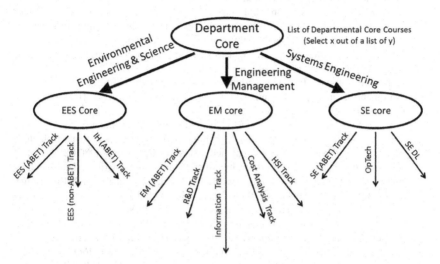

Consolidated Program Design, Evaluation, Justification, and Integration

FIGURE 7.5   Case example for curriculum integration.

NOTES

1 https://www.theguardian.com/books/2011/sep/22/william-gibson-beyond-cyber space

2 https://www.theguardian.com/books/2011/sep/22/william-gibson-beyond-cyberspace. "Cyberspace is a realm of total-immersion virtual reality," just as when Case, the hero of Neuromancer (1984), applies the dermatrodes of his cyberspace deck to his forehead, powers it up and jacks in to the matrix, his "inner eye" sees as "transparent 3D chessboard extending to infinity," on which, or in which, is "a graphic representation of data abstracted from the banks of every computer in the human system."

3 https://www.businessinsider.com/solarwinds-hack-explained-government-agencies-cyber-security-2020-12

4 https://www.bloomberg.com/news/articles/2021-06-04/hackers-breached-colonial-pipeline-using-compromised-password

5 Badiru, Adedeji B.; *Quality Insights: The DEJI Model for Quality Design, Evaluation, Justification, and Integration*; Int. J. Quality Engineering and Technology, Vol. 4 No. 4, 2014.

6 Warren and Brandeis, "The Right to Privacy" Harvard Law Review, Vol. IV, December 15, 1890.

7 Section 224 refers to these operational security standards as trusted supply chain and operational security standards and requires the Secretary of Defense (SECDEF) to establish these security standards no later than January 1, 2021.

8 Cyber physical systems are defined as technologies that combine the cyber and physical worlds that can respond in real time to their environments. Cyber physical systems include electronic parts, assemblies, systems, and system elements that

operate as a single, self-contained device or within an interconnected network providing shared operations. An added distinction of this CPS definition is a requirement affecting a tangible physical output through command and control embedded in the device or distributed across network nodes. The command and control system could be manipulated through a physical process and/or physical input of the data flow (e.g., manipulating sensor input to spoof or disrupt the expected data). Draft SAE JA7496 (December 2020).

9   Hardware, software, firmware, and cyber physical systems (CPS).

10 Argument-based decision support for risk analysis; Sven Ove Hanssona* and Gertrude Hirsch Hadornb; 15 February 2017

11 https://www.nasa.gov/consortium/ModelBasedSystems; November 5, 2021.

12 Test and evaluation master plan is a performance assessment plan that will include schedule, test types and environment, and assets required; DoDI 5000.02.

13 Use of a quantitative data report in a hypothetical decision scenario for health policy-making: a computer-assisted laboratory study; Pamela Wronski1*, Michel Wensing1, Sucheta Ghosh2, Lukas Gärttner1, Wolfgang Müller2and Jan Koetsenruijter1.

14 Use of a quantitative data report in a hypothetical decision scenario for health policy-making: a computer-assisted laboratory study; Pamela Wronski1*, Michel Wensing1, Sucheta Ghosh2, Lukas Gärttner1, Wolfgang Müller2and Jan Koetsenruijte.

15 Use of a quantitative data report in a hypothetical decision scenario for health policy-making: a computer-assisted laboratory study; Pamela Wronski1*, Michel Wensing1, Sucheta Ghosh2, Lukas Gärttner1, Wolfgang Müller2and Jan Koetsenruijter1.

16 Argument-based decision support for risk analysis; Sven Ove Hanssona* and Gertrude Hirsch Hadornb; 15 February 2017.

17 Use of a quantitative data report in a hypothetical decision scenario for health policy-making: a computer-assisted laboratory study; Pamela Wronski1*, Michel Wensing1, Sucheta Ghosh2, Lukas Gärttner1, Wolfgang Müller2and Jan Koetsenruijter1.

18 Argument-based decision support for risk analysis; Sven Ove Hanssona* and Gertrude Hirsch Hadornb; 15 February 2017.

19 Argument-based decision support for risk analysis; Sven Ove Hanssona* and Gertrude Hirsch Hadornb; 15 February 2017.

20 Shutdown Maintenance Drivers under an Integrated and Business Focused Maintenance System; Peter Muganyi and Charles Mbohwa.

21 Argument-based decision support for risk analysis; Sven Ove Hanssona* and Gertrude Hirsch Hadornb; 15 February 2017.

22 Bommer, John S.; Understanding Cyber Risk Assessment, October 2021.

23 CIKR "includes those assets, systems, networks, and functions—physical or virtual—so vital to the United States that their incapacitation or destruction would have a debilitating impact on security, national economic security, public health or safety, or any combination of those matters." Critical Infrastructure and Key Resources Support Annex, January 2008.

24 Young & Leveson (2014), "An Integrated Approach to Safety and Security Based on Systems Theory, Communications of the ACM," Vol 57, No 2.

25 Critical Infrastructure and Key Resources Support Annex, p. CIKR-6, January 2008.

26 J.A. Zachman, "A Framework for Information Systems Architecture," *IBM Systems Journal* 26, vol. 26, no. 3, pp. 276–292, 1987.

27 https://en.wikipedia.org/wiki/Perimeter_security

28 Security perimeter is refers to natural barriers or built fortifications to either keep intruders out or to keep captives contained within the area the boundary surrounds.

29 Bommer, John S.; Cybersecurity for Autonomous Vehicles—Cyber Risk Assessment; University of Dayton Manufacturing Day, October 2019.

30 Firmware is the computer programs and data stored in the hardware such that the programs and data cannot be dynamically written or modified during the execution. However, firmware has an embedded, programmable logic that operates in a read-only modes that actives and runs. This software is susceptible to supply chain attack during the building and configuration of the firmware SBOM creation and maintenance. Reference: SAE International Draft JA7496, December 2020.

31 Provenance is the chain of custody identifying each person or organization handling the component and provides insight as to how it was handled or processed throughout the component life cycle. This includes after it leaves the control of its developer(s) and enters the supply chain. Provenance also reflects changes in responsibility for the ongoing development of the component (new versions, patches, etc.). For example, if this responsibility shifts from the original developer to an integrator or a new development organization (as when one organization acquires another). Reference: SAE International Draft JA7496, December 2020.

32 Pedigree is the compositional genetics (background/lineage) of "what" is in the component under consideration. This includes such considerations as the version of the component at a given point in time, including when it was originally conceived, implemented, composed, or modified. Reference: SAE International Draft JA7496, December 2020.

33 Bill of Materials (BOM) is the structure or list of all the parts needed to build the end product. Parts include discrete components, sub-assemblies, raw materials, software components, sub-assemblies, firmware raw materials, firmware and labor needed to build the CPS, software, or firmware of the end product. A BOM can be hierarchical or flat. A hierarchical BOM displays a multi-level view of the software, sub-assemblies and reusable blocks, used to build the product, whereas a flat BOM, displays a consolidated list of every line of code or the representative unique item data structure.

34 Bommer, John S.; Artis, James; Bauer, Parker; and Wright, Alexander; Supply Chain Risk Management and Cyber-Supply Chain Risk Management; AFIT/LS Supply Chain Risk Management Workshop; April 2021.

35 Plappert, Christian; Daniel, Elle; Gadacz, Henry; Roland, Rieke; Scheuermann, Dirk; Krauss, Christoph; *Attack Surface Assessment for Cybersecurity Engineering in the Automotive Domain*; 2021 29th Euromicro International Conference on Parallel, Distributed and Network-Based Processing.

36 Lykou, Georgia; Anagnostopoulou, Argiro; and Gritzalis, Dimitris; *Smart Airport Cybersecurity: Threat Mitigation and Cyber Resilience Controls;* The 2018 IEEE Global IoT Summit (GIoTS), Bilbao, Spain, 4–7 June 2018.

37 Badiru, Adedeji B.; Application of the DEJI Model for Aerospace Product Integration; Air Force Institute of Technology (AFIT) Wright Patterson Air Force Base, Ohio.

38 Bommer, John S. and Artis, James; Supply Chain Risk Management and Cyber-Supply Chain Risk Management; AFIT/LS Supply Chain Risk Management Workshop; April 2021.

8 Data Modeling for Systems Integration

INTRODUCTION

Business modeling and forecasting are useful to adapt to changes in a business process and for optimizing business process management. DEJI Systems Model® is amenable to the application of quantitative data modeling in the design, evaluation, justification, and integration stages. Data-driven decision-making eliminates guesswork, hypothesis, and corporate politics from decision-making. This improves the business performance by highlighting the areas that have the maximum impact on the operational efficiency and revenues. Businesses today are overloaded with data on finances, operations, and customer purchases. Increasingly, executives are now leaning on data analytics to make informed business decisions, thus eliminating the intuition and gut feel. Modeling can bring a scientific angle to the management of any businesses. By reducing the tremendous amount of raw data into actionable information, modeling leads the way to smarter and more accurate decisions

A chapter in Badiru et al. (2019) presents case examples of considerations and systems approaches for business systems modeling on an enterprise-wide scale. That chapter is adapted for the case study presented here. This chapter brings about modeling and decision automation to improve the quality and reliability of data for efficient decision-making. It consists of advanced tools and expert systems to sharpen business operations and contribute to competitive advancement. Modeling is integral to business management. Models are frameworks to assess competing courses of action. They help compare risk-reward scenarios and test ideas. Modeling is also pervasive. Every time spreadsheets are built to run the numbers, models are being built. Modeling is at the core of enterprise strategy and treating it as such is nothing less than imperative. Applying modeling techniques on an enterprise scale is frequently called enterprise simulation (ES). It has two distinct objectives:

- Provide a top-down view of the business enterprise to support strategic decision-making.
- Enable workers across the enterprise with modeling tools and techniques for routine decisions.

Business intelligence (BI) and similar historical reporting systems help monitor business operations, but they cannot evaluate new ideas for which there is no history. Spreadsheets are flexible but unable to scale or foster collaboration. Business analytics modeling is unique in its ability to satisfy ES objectives, to cut modeling

DOI: 10.1201/9781003175797-8

costs by simplifying model structures and encouraging reuse of prior work, and to improve model accuracy by dividing tasks among those with the best domain knowledge.

The ability to model and perform decision modeling and analysis is an essential feature of many real-world applications ranging from emergency medical treatment in intensive care units to military command and control systems. The analysis and modeling of real-world problems are important features for performing decision-making in diverse applications. This ranges from manufacturing applications, construction, military applications, health applications, logistic, transportation distribution, to mention just a few. Models are essential in providing support for businesses processes and systems and dealing with complex problems. The development of appropriate models for planning and management is a tool for improving efficiency in real-world problems. Consequently, models are developed, with the aim of computing estimates and forecasts for real-world data. Modeling helps make informed decisions, using techniques for analysis, estimation, and forecasting. With modeling, data can be used to describe realities, build scenarios, and predict performances.

Almost all decision-making is based on forecasts. Every decision becomes operational at some point in the future, so it should be based on forecasts of future conditions. Forecasts are needed continually, and as time moves on, the impact of the forecasts on actual performance is measured; original forecasts are updated; and decisions are modified, and so on. If models are accurate, reliable forecasts can be computed based on the models. The resulting forecasts yield the same performances as that of the models.

Almost all managerial decisions are based on forecasts. Every decision becomes operational at some point in the future, so it should be based on forecasts of future conditions. Forecasts are needed throughout an organization, and they should certainly not be produced by an isolated group of forecasters. Forecasts are needed continually, and as time moves on, the impact of the forecasts on actual performance is measured; original forecasts are updated; and decisions are modified, and so on. The decision-maker uses forecasting models to assist him or her in decision-making process. The decision-maker often uses the modeling process to investigate the impact of different courses of action under a course of action.

MODELING THEORY

A model is defined as a representation of a system for the purpose of studying the system. A model is a representation and abstraction of anything such as a real system, a proposed system, a futuristic system design, an entity, a phenomenon, or an idea. It is necessary to consider only those aspects of the system that affect the problem under investigation. These aspects are represented in a model, and by definition it is a simplification of the system. It is possible to derive a model based on physical laws, which makes it possible to calculate the value of some time-dependent quantity nearly exactly at any instant of time. If exact calculation were possible, such a model would be entirely *deterministic*. Nevertheless, it may be possible to derive a model that can be used to calculate the probability of a future behavior of the value lying between two specified limits. Such a model is called a probability model or a *stochastic model.*

The external components which interact with the system and produce necessary changes are said to constitute the system environment. In modeling systems, it is necessary to decide on the boundary between the system and its environment. This decision may depend on the purpose of the study. A model is closely related to the system. A system is defined as an aggregation or assemblage of objects joined in some regular interaction or interdependence toward the accomplishment of some purpose. The system environment constitutes external components which interact with the system and produce necessary changes to the system. In modeling systems, it is necessary to decide on the boundary between the system and its environment. This decision may depend on the purpose of the study. The components of a system are defined as follows:

Entity: An entity is an object of interest in a system. For example, in a production system, departments, orders, parts, and products are the entities.

Attribute: An attribute denotes the property of an entity. For example, quantities for each order, types of parts, or number of machines in a department are attributes of a production system.

Activity: Any process causing changes in a system is called an activity. For example, the fabrication process of a department.

State of a system: The state of a system is defined as the collection of variables necessary to describe a system at any time, relative to the objective of study. In other words, state of the system means a description of all the entities, attributes, and activities as they exist at one point in time.

Example: In a production system, the arrival of orders may be considered to be outside the company but yet a part of the system environment. There is a relationship between company output and arrival of orders in considering demand and supply of goods. This relationship is considered an activity of the system.

Every study begins with a statement of the problem, provided by policy makers. Analysts ensure it's clearly understood. This step is referred to as problem formulation. Another step of the modeling process is the setting up of objectives. The objectives indicate the questions to be answered by identifiable problems. The construction of a model of a system or model conceptualization is an art as well as a science. The modeling process is characterized by the ability (i) to abstract the essential features of a problem (ii) to select and modify basic assumptions that characterize the system, and (iii) to enrich and elaborate the model until a useful approximation results are obtained. The model building process enhances the quality of the resulting model and increases the confidence of the model user in the application of the model. Real-world systems result in models that require a great deal of information storage and computation. These models can be programmed using system languages or special-purpose software. Examples of these software include MATLAB®, Modeller, MacA&D, ArchCAD,

AutoCAD, and CATIA. The programming of these models is referred to as the model translation.

It is pertinent to verify and validate the models by checking the performance of the models developed to represent real-world situations. If the input parameters and logical structure are correctly represented, verification is completed. As part of the calibration process of a model, the modeler must validate and verify the model. The term "validation" is applied to those processes which seek to determine whether or not a model is correct with respect to the "real" system. More prosaically, validation is concerned with the question "Are we building the right system?" Verification, on the other hand, seeks to answer the question "Are we building the system right?". The purpose of modeling is to determine that a model is an accurate representation of the real system. This is achieved through calibration of the model, which is an iterative process of comparing the model to actual system behavior, and the discrepancies between the two are determined.

The model obtained for a problem can be generic, that is, can be used again by the same or different analysts to solve different problems. Further modifications can be made. Model users can change the input parameters for better performance. Success of the model depends on if model requirements and outputs are fully implemented. The purpose of models is to aid in designing solutions. They are to assist with understanding of the problem and to aid deliberation and choice by allowing us to evaluate the consequence of our actions before implementing them.

TYPES OF MODELS

In modeling theory, there are many models that are designed but a few of them are used. In formulating models, the concept of "implementation" is defined, and we progressively shift from a traditional "design then implementation" standpoint to a more general theory of a model design/implementation, seen as a cross-construction process between the model and the organization in which it is implemented. Consequently, the organization is considered not as a simple context but as an active component in the design of models. In model-based decision-making, we are particularly interested in the idea that a model is designed with a view to action. Basically, there are various types of models. They are:

Mathematical Model: This is one in which symbols and logic constitute the model. The symbolism used can be language or mathematical notations. This type of model uses mathematical equations to represent a system.

Example: A typical example of a mathematical model is a simulation model. This is the study of the behavior of a system as it evolves over time. This model takes the form of a set of assumptions concerning the operation of the system. These assumptions are expressed in the form of

- Mathematical relationships
- Logical relationships
- Symbolic relationships

Physical (Iconic) Model: Physical model is a smaller or larger physical copy of an object.

The object being modeled may be small (for example, an atom) or large (for example, the solar system). Other examples include a map, a globe, or a model car.

Static Model: This method is also known as the Monte-Carlo method. This system represents a system at a particular point of time. It describes relationships that do not change with respect to time. Examples of this type of model include an architectural model of a house, or an equation relating the lengths and weights on each side of a playground variation.

Dynamic Model: This type of model represents systems as they change over time. The dynamic model describes time-varying relationships. Examples include a wind tunnel, equations of motion of the planets around the sun, which constitute a dynamic model of the solar system. Dynamic system means a system capable of action and/or change.

Deterministic Model: This type of model contains no random variables. They have a known set of inputs, which will result in a unique set of outputs. An example is the arrival of patients at a hospital at their scheduled appointment time.

Stochastic Model: The stochastic model has one or more random variables as inputs. These random inputs lead to random outputs. For example, a banking system involves random inter-arrival and service times.

Discrete Model: This is the discrete analog of continuous modeling. In discrete models, formulae are fit to discrete data. Discrete data are data that could potentially take only a countable set of values. These could be integers and are not infinitely divisible. An example is the number of cells in a population, displayed at regular time intervals.

Continuous Model: This is the mathematical practice of applying a model to continuous data (data which has an infinite number and divisibility of attributes). They often are in the form of differential equations and are converted to discrete models. An example is the amount of water in a tank and or its temperature.

There are guidelines for an analyst to successfully implement a model that could be both valid and legitimate. These guidelines are as follows:

1. The analyst must be ready to work in close cooperation with the strategic stakeholders in order to acquire a sound understanding of the organizational context. In addition, the analyst should constantly try to discern the kernel of organizational values from its more contingent part.
2. The analyst should attempt to strike a balance between the level of model sophistication/complexity and the competence level of stakeholders. The model must be adapted both to the task at hand and to the cognitive capacity of the stakeholders.

3. The analyst should attempt to become familiar with the various preferences prevailing in the organization. This is important since the interpretation and the use of the model will vary according to the dominant preferences of the various organizational actors.

4. The analyst should make sure that the possible instrumental uses of the model are well documented and that the strategic stakeholders of the decision-making process are quite knowledgeable about and comfortable with the contents and the working of the model.

5. The analyst should be prepared to modify or develop a new version of the model, or even a completely new model, if needed, that allows an adequate exploration of heretofore unforeseen problem formulation and solution alternatives.

6. The analyst should make sure that the model developed provides a buffer or leaves room for the stakeholders to adjust and readjust themselves to the situation created by the use of the model.

7. The analyst should be aware of the pre-conceived ideas and concepts of the stakeholders regarding problem definition and likely solutions; many decisions in this respect might have been taken implicitly long before they become explicit.

MODELING FOR FORECASTING

Forecasting is an important tool that is useful in planning, whether in business or government. Often, forecasts are generated subjectively and at great cost by group discussion, even when relatively simple quantitative methods can perform just as well or, at the very least; provide an informed input to such discussions. The modeling process is useful for:

1. Understanding the underlying mechanism generating the time series. This includes describing and explaining any variations, seasonality, trend, etc.
2. Predicting the future
3. Controlling the system

The selection and implementation of proper forecasting models is necessary for planning purposes in organizations. Usually, the financial well-being of an organization relies on the accuracy of the forecast since such information will likely be used to make interrelated budgetary and operative decisions in areas of personnel management, purchasing, marketing and advertising, capital financing, etc. For example, under-forecasts may cause an organization to be overly burdened with excess assets costs, and hence create lost sales revenue through unanticipated item shortages. There are two main approaches to forecasting. Either the estimates of future values are based on the analysis of factors which are believed to influence future values, that is, the explanatory method, or else the prediction is based on an inferred study of past general data behavior over time, that is, the extrapolation method. For example, the belief that the sale of doll clothing will increase from current levels because of a recent advertising blitz rather than proximity to Christmas illustrates the difference between the two philosophies. It is possible that both approaches will lead to the creation of accurate and useful forecasts, but it must be remembered that, even for a modest degree of desired accuracy, the former method is often more difficult to implement and validate than the latter approach.

TIME-SERIES FORECASTING

Organizations with large operation and staff comprising of statisticians, management scientists, computer analysts, etc. are in a much better position to select and make proper use of sophisticated forecast techniques than companies with more limited resources. Notably, the bigger firm, through its larger resources, has a competitive edge over smaller organizations and can be expected to be very diligent and detailed in estimating forecasts.

A time series is a sequence of numerical data points in successive order. Generally, a time series is a sequence taken at successive equally spaced points in time. It is a sequence of discrete-time data. For example, in investment analysis, a time-series tracks the movement of the chosen data points, such as a security's price over a specified period of time with data points recorded at regular intervals. One of the main goals of time-series analysis is to forecast future values of the series. A trend is a regular, slowly evolving change in the series level. Changes that can be modeled by low-order polynomials. There are three general classes of models that can be constructed for purposes of forecasting or policy analysis. Each involves a different degree of model complexity and presumes a different level of comprehension about the processes one is trying to model. In making a forecast, it is also important to provide a measure of how accurate one can expect the forecast to be. The use of intuitive methods usually precludes any quantitative measure of confidence in the resulting forecast. The statistical analysis of the individual relationships that make up a model, and of the model as a whole, makes it possible to attach a measure of confidence to the model's forecasts.

In time-series models, we presume to know nothing about the causality that affects the variable we are trying to forecast. Instead, we examine the past behavior of a time series in order to infer something about its future behavior. The method used to produce a forecast may involve the use of a simple deterministic model such as a linear extrapolation or the use of a complex stochastic model for adaptive forecasting. One example of the use of time-series analysis would be the simple extrapolation of a past trend in predicting population growth. Another example would be the development of a complex linear stochastic model for passenger loads on an airline. Time-series models have been used to forecast the demand for airline capacity, seasonal telephone demand, the movement of short-term interest rates, and other economic variables. Time-series models are particularly useful when little is known about the underlying process one is trying to forecast. The limited structure in time-series models makes them reliable only in the short run, but they are nonetheless rather useful.

In regression models, the variable under study is explained by a single function (linear or nonlinear) of a number of explanatory variables. The equation will often be time-dependent (i.e., the time index will appear explicitly in the model), so that one can predict the response over time of the variable under study. The main purpose for constructing regression models is forecasting. A forecast is a quantitative estimate (or set of estimates) about the likelihood of future events which is developed on the basis of past and current information. This information is embodied in the form of a model. This model can be in the form of a single-equation structural model, a multi-equation model, or a time-series model. By extrapolating our models beyond the period over which they were estimated, we can make forecasts about near-future events.

The term "forecasting" is often thought to apply solely to problems in which we predict the future. An example of a single-equation regression model would be an equation that relates a particular interest rate, such as the money supply, the rate of inflation, and the rate of change in the gross national product. The choice of the type of model to develop involves trade-offs between time, energy, costs, and desired forecast precision. The construction of a multi-equation simulation model may require large expenditures of time and money. The gains from this effort may include a better understanding of the relationships and structure involved as well as the ability to make a better forecast. However, in some cases these gains may be small enough to be outweighed by the heavy costs involved. Because the multi-equation model necessitates a good deal of knowledge about the process being studied, the construction of such models may be extremely difficult.

The decision to build a time-series model usually occurs when little or nothing is known about the determinants of the variable being studied, when a large number of data points are available, and when the model is to be used largely for short-term forecasting. Given some information about the processes involved, however, it may be reasonable for a forecaster to construct both types of models and compare their relative performance. Two types of forecasts can be useful. Point forecasts predict a single number in each forecast period, while interval forecasts indicate an interval in which we hope the realized value will lie.

VERIFICATION AND VALIDATION OF MODELS

As a means of verification and validation of estimates obtained from a model, the estimates are compared with actual data. A good model should have small error measures between the estimated values and actual data. A model is acceptable if the *mean average percentage error* (*MAPE*) of the model is less than 20%. This validation technique investigates the performance of newly developed models compared with actual data. The process of computing: *MAPE* values are outlined below.

The forecast errors are computed from a time series, based on an average of weighted past observations. At period t, past values of a variable of interest X_t can be observed or values forward in the future. The model is applied to the historical observations, and forecasted values F_{t+1} are obtained. F_{t+1} are values achieved as a result of making predictions of the future based on past and present data. To identify an accurate predictive model, the following steps are followed:

Choose a forecasting method based on the observed pattern of the time series.
Use the forecasting method to develop fitted values of the data.
Calculate the forecast error.
Make a decision about the appropriateness of the model based on the measure of forecast error.

For the purpose of computing forecasting errors, a historical data set called a time series is considered. A time series consists of the data of interest on a single variable which has been collected in a consistent way over time at equally spaced intervals. Time series are analyzed to search for patterns by graphing or plotting time-series data set. The one-sided moving average of past n observations is given as:

$$F_{t+1} = \frac{X_t + X_{t-1} + \ldots + X_{t-n+1}}{n}$$

$$= \frac{1}{n}\left(\sum_{i=t-n+1}^{t} X_i\right)$$

where t is the most recent observation and $t + 1$ is the next period. This formula requires that the values of the past n observations are known. Accordingly, the concept of adding a new observation and dropping the oldest observation, the formula is restated as

$$F_{t+1} = \frac{1}{n}\left(\sum_{i=t-n}^{t-1} X_i\right) + \frac{1}{n}\left(X_t - X_{t-n}\right)$$

$$= F_t + \frac{X_t}{n} - \frac{X_{t-n}}{n}$$

The formula representing the time series is an adjustment of the forecast F_t in the previous period. If n is increased, a much smaller adjustment is made for each new time period.

This representation of time series can be developed, by making the substitution $F_t = X_{t-n}^2$, to get

$$F_{t+1} = \frac{X_t}{n} - \frac{F_t}{n} + F_t$$

Furthermore, this can then be rewritten as

$$F_{t+1} = \frac{1}{n} X_t + \left(1 - \frac{1}{n}\right) F_t$$

This is a forecast based on weighting the most recent observations with a weight of value $1/n$ and weighting the most recent forecast with a weight of $1 - 1/n$. Since the number of periods n is a constant, the fraction $1/n$ must be greater than zero and less than unity. If $\propto$ is substituted for $1/n$, the basic model is written as

$$F_{t+1} = \propto X_t + \left(1 - \propto\right) F_t$$

where t is the current time period, F_{t+1} and F_t are the forecast values for the next and current periods, and X_t is the current observed value. α is called the smoothing constant and it takes values between zero and unity. If F_t is expressed in terms of the preceding observed T_{t-1} and the forecast F_{t-1} values, then the equivalent for the next period's forecast becomes

$$F_{t+1} = \propto X_t + \left(1 - \propto\right)\left[\propto X_{t-1} + \left(1 - \propto\right) F_{t-1}\right]$$

which simplifies to

$$F_{t+1} = \propto X_t + \propto \left(1 - \propto\right) X_{t-1} + \left(1 - \propto\right)^2 F_{t-1}$$

This can continue for several earlier periods, which show that all preceding values of X are reflected in the current forecasts. Thus, the successive weights $\propto$, $\propto (1 - \propto)$, $(1 - \propto)^2$, ... decrease exponentially. This can be rewritten as follows:

$$F_{t+1} = F_t + \propto \left(X_t - F_t\right)$$

and this is simply

$$F_{t+1} = F_t + \propto e_t$$

where the forecast error e_t for period t is just the actual minus the forecast. Thus, the forecast provided by the time series is the old forecast plus an adjustment for the error occurring in the last forecast.

To evaluate the accuracy of a predictive technique, it is required to evaluate the error in forecasting. The error in period t was defined as the actual value X_t minus the predicted value F_t:

$$e_t = X_t - F_t$$

An examination of the error in forecasting permits the evaluation of whether the chosen prediction model accurately mirrors the pattern exhibited in the sample observations. An evaluation of the reliability of a model requires the specification of criteria.

The MAPE is based on the assumption that the severity of error is linearly related to its size. The MAPE is the sum of the absolute values of the errors divided by the corresponding observed values divided by the number of forecasts. The MAPE is often expressed as a percentage. This is to provide meaningful interpretation about each data point. The MAPE is defined as:

$$\text{MAPE} = \frac{\sum |e_t|/X_t}{n} \times 100$$

If the predictive model is accurate, reliable forecasts can be computed based on the predictive model. Future demand for power can be determined by developing accurate estimates for predictive models. Predictive models are static, constructed from historical data. The resulting forecasts yield the same performance as that of the predictive model.

The mean average percentage error (MAPE) is given as:

$$\text{MAPE} = \frac{\sum |e_t|/X_t}{n} \times 100$$

where

$$e_t = X_t - F_t$$

$$\text{MAPE} = \frac{\sum |e_t|/X_t}{n} \times 100 = \frac{2.762923412}{23} \times 100 = 12.0127\%$$

Since the MAPE is less than 20%, the estimated values for the model are acceptable.

FORECASTING TECHNIQUES

Forecasting is a prediction of what will occur in the future, and it is an uncertain process. Because of the uncertainty, the accuracy of a forecast is as important as the outcome predicted by the forecast. This section presents a general overview of business forecasting techniques. The forecasting techniques considered in this section include regression analysis, Box-Jenkins, artificial neural network, and Kalman techniques.

Some of the aforementioned techniques can be solved analytically, for example the regression technique and the moving averages. The analytical method is used to solve a specific issue. This is a procedure for the analysis of some problem or fact. However, some very complex problems are solved using simulation method, for example the Box-Jenkins technique and the artificial neural network. This is the imitation of a real-world process. The act of simulation requires that a model be developed. The model represents the key characteristics, behaviors, and functions of the process.

REGRESSION ANALYSIS

Regression techniques belong to the class of causal models. Regression is the study of relationships among variables, a principal purpose of which is to predict, or estimate the value of one variable from known or assumed values of other variables related to it. To make predictions or estimates, we must identify the effective predictors of the variable of interest. To develop a regression model, begin with a hypothesis about how several variables might be related to another variable and the form of the relationship. A regression using only one predictor is called a simple regression. Where there are two or more predictors, multiple regressions analysis is employed.

$$\text{Given a line } y = mx + b$$

We determine what the best line's slope (m) and intercept (b) should be for a given set of data (x,y). We could randomly guess the m and b values of the line, tabulate the error of each, and then identify which guess results in the least error. The given equation can be estimated as

$$y = \overset{\frown}{m}x + \overset{\frown}{b}$$

This method is called a least squares linear regression method. We first minimize the error. The derivative of the error function reaches a minimum and its slope is equal to zero when the error value is also at a minimum. Let

$$\varepsilon^2 = \text{error}^2 = \Sigma\left(y_i - y_p\right)^2$$

and,

$$y_p = m\,x_i + b + \varepsilon$$

where y_p = predicted value when $x = x_i$, and y_i = experimental value when $x = x_i$.
Then as before,

$$\varepsilon^2 = \Sigma\left(y_i - m\,x_i - b\right)^2$$

$$\varepsilon^2 = \Sigma\left(m^2 x_i^2 + 2mbx_i - 2mx_iy_i + b^2 - 2by_i + y_i^2\right)$$

As we approach the best value for m and b, ε^2 approaches its minimum value where the error as a function of m and b is at a minimum. Solve (*) for the minimum as a function of m. We seek the point where the first-derivative equals zero

$$\frac{d\varepsilon^2}{dm} = 0,$$

$$\frac{d\varepsilon^2}{dm} = 2m\Sigma xi2 + 2b\Sigma xi - 2\Sigma(xiyi) = 0$$

$$2m\Sigma x_i^2 + 2b\Sigma x_i - 2\Sigma(x_iy_i) = 0$$

$$2b\Sigma xi = 2\Sigma(xiyi) - 2m\Sigma xi2$$

$$b\Sigma xi = \Sigma(xiyi) - m\Sigma xi2$$

$$b = \frac{\Sigma(xiyi)}{\Sigma xi} - \frac{m\Sigma xi^2}{\Sigma xi}$$

$$b = \frac{\Sigma(xiyi) - m\Sigma xi^2}{\Sigma xi}$$

Solve the equation for the minimum as a function of b and seek the point where the first-derivative equals zero.

$$\frac{d\varepsilon^2}{db} = 0$$

$$\frac{d\varepsilon^2}{db} = 2m\Sigma xi + 2\Sigma b - 2\Sigma yi = 0$$

$$2m\Sigma xi + 2\Sigma b - 2\Sigma yi = 0$$

$$2\Sigma b = 2\Sigma y_i - 2m\Sigma xi$$

To simplify the notation (for number of points $= n$), let

$$Sx = \Sigma x_i, \ Sy = \Sigma y_i, \ Sxy = \Sigma(x_iy_i), \ Sxx = \Sigma(x_i^2), \ nb = \Sigma b$$

From [3.13].

$$mx_i = y_p - b$$

From the above steps, solve for m:

$$m = Sxy - bSxSxx$$

Or, stated differently, $Sxy = mSxx + bSx$.

BOX JENKINS TECHNIQUE

The Box-Jenkins technique (Box et al., 2008) consists of a family of time-series models. It comprises many different models. These models can be grouped into three basic classes—autoregressive models, moving average models, and autoregressive-integrated moving average models. For many problems in business, engineering, and physical and environmental sciences, Box-Jenkins technique may be applied to related variables of interest. An analysis may be obtained by considering individual series as components of a multivariate or vector time series and analyzing the series jointly. The Box-Jenkins technique is used to study the relationship among variables. This involves the development of statistical models and methods of analysis that adequately describe the inter-relationships among the series.

The models for time series are stochastic models. A time series $z_1, z_2, ..., z_n$ of N successive observations is regarded as a sample realization from an infinite population of such time series that could have been generated by the stochastic process. The *backward shift operator B* is defined by $Bz_t = z_{t-1}$; hence $B^m z_t = z_{t-m}$. Another important operator is the *backward difference operator*, ∇, defined by $\nabla z_t = z_t - z_{t-1}$. This can be written in terms of B, since

$$\nabla z_t = z_t - z_{t-1} = (1 - B) z_t$$

The stochastic models employed are based on the idea that an observable time series z_t in which successive values are highly dependent can frequently be regarded as generated from a series of independent "shocks" a_t. These *shocks* are random drawings from a fixed distribution, usually assumed normal and having mean zero and variance σ_a^2. Such a sequence of independent random variables a_t, $a_{t-1}, a_{t-2}...$ is called a *white noise* process. The white noise process a_t is supposedly transformed into the process z_t by what is called a linear filter. The linear filtering operation simply takes a weighted sum of previous random shocks a_t, so that

$$z_t = \mu + a_t + \psi_1 a_{t-1} + \psi_2 a_{t-2} + ...$$

$$= \mu + \psi(B) a_t$$

In general, μ is a parameter that determines the "level" of the process, and

$$\psi(B) = 1 + \psi_1 B + \psi_2 B^2 + ...$$

is the linear operator that transforms a_t into z_t, and is called the transfer function of the filter. The model representation of equations given above can allow for a flexible range of pattern of dependence among values of the process $\{z_t\}$ expressed in terms of the independent random shocks a_t. The sequence $\psi_1, \psi_2, \ldots$ formed by the weights may, theoretically, be finite or infinite. If the sequence is finite or infinite and absolutely summable in the sense that

$\sum_{j=0}^{\infty} |\psi_j| < \infty$, the filter is said to be stable and the process z_t is stationary. The parameter μ is then the mean about which the process varies. Otherwise, z_t is nonstationary and μ has no specific meaning except as a reference point for the level of the process.

Denote the values of a process at equally spaced times $t, t-1, t-2 \ldots$ by z_{t-1}, $z_{t-2}, \ldots$.

Also let $\tilde{z}_t = z_t - \mu$ be the series of deviations from μ. Then

$$\tilde{z}_t = \phi_1 \tilde{z}_{t-1} + \phi_2 \tilde{z}_{t-2} + \ldots + \phi_p \tilde{z}_{t-p} + a_t$$

is called an *autoregressive* (AR) process of order p. The reason for this name is that a linear model

$$\tilde{z} = \phi_1 \tilde{x}_1 + \phi_2 \tilde{x}_2 + \ldots + \phi_p \tilde{x}_p + a$$

relating a "dependent" variable z to a set of "independent" variables $x_1, x_2, \ldots x_p$, plus a random error term a, is referred to as a regression model, and z is said to be "regressed" on previous values of itself: hence the model is *autoregressive*. If an *autoregressive operator* of order p is defined in terms of the backward shift operator B by

$$\phi(B) = 1 - \phi_1 B - \phi_2 B^2 - \ldots - \phi_p B^p,$$

The autoregressive model given may be written as

$$\phi(B) \tilde{z}_t = a_t$$

The model contains $p+2$ unknown parameters, $\phi_1, \phi_2, \ldots \phi_p, \sigma_a^2$, which in practice have to be estimated from the data. The additional parameter σ_a^2 is the variance of the white noise process a_t.

The autoregressive (AR) model is a special case of the linear filter model. For example, $\tilde{z}_{t-1}$ can be eliminated from the right-hand side of the autoregressive model by substituting

$$\tilde{z}_{t-1} = \phi_1 \tilde{z}_{t-2} + \phi_2 \tilde{z}_{t-3} + \ldots + \phi_p \tilde{z}_{t-p-1} + a_{t-1}$$

Similarly, $\tilde{z}_{t-2}$ can be substituted, and so on, to yield eventually an infinite series in the a's. Consider, specifically, the simple first-order ($p = 1$) AR process,

$\tilde{z}_t = \phi\tilde{z}_{t-1} + a_t$. After m successive substitutions of $\tilde{z}_{t-1} = \phi\tilde{z}_{t-1} + a_{t-j,j=1,...,m}$, in the right-hand side we obtain

$$\tilde{z}_t = \phi^{m+1}\tilde{z}_{t-m-1} + a_t + \phi a_{t-1} + \phi^2 a_{t-2} + ... + \phi^m a_{t-m}$$

In the limit as $m \to \infty$, this leads to the convergent infinite series representation $\tilde{z}_t = \sum_{j=0}^{\infty} \phi^j a_{t-j}$ with $\psi_j = \phi^j, j \geq 1$, provided that $|\phi| < 1$. Symbolically, in the general AR case, we have that

$$\phi(B)\tilde{z}_t = a_t$$

is equivalent to

$$\tilde{z}_t = \phi^{-1}(B)a_t = \psi(B)a_t$$

with $\psi(B) = \phi^{-1}(B) = \sum_{j=0}^{\infty} \psi_j B^j$.

Autoregressive processes can be stationary or nonstationary. For the process to be stationary, the ϕ's must be such that the weights ψ_1, ψ_2, ... in $\psi(B) = \phi^{-1}(B)$ form a convergent series. The necessary requirement for stationarity is that the autoregressive operator, $(B) = 1 - \phi_1 B - \phi_2 B^2 - ... - \phi_p B^p$, considered as a polynomial in B of degree p, must have all roots of $\phi(B) = 0$ greater than 1 in absolute value; that is, all roots must lie outside the unit circle. For the first-order AR process $\tilde{z}_t = \phi\tilde{z}_{t-1} + a_t$, this condition reduces the requirement that $|\phi| < 1$, as the argument above has already indicated.

The general form of the model that is used to describe time series is the ARIMA model

$$\varphi(B)z_t = \phi(B)\nabla^d z_t = \theta_0 + \theta(B)a_t$$

where

$$\phi(B) = 1 - \phi_1 B - \phi_2 B^2 - ... - \phi_p B^p$$

$$\theta(B) = 1 - \theta_1 B - \theta_2 B^2 - ... - \theta_q B^q$$

$\phi(B)$ and $\theta(B)$ are polynomial operators in B of degrees p and q. This process is referred to as an ARMA(p, q) (Auto Regressive Moving Average) process. The ARIMA (Auto Regressive Integrated Moving Average) model can be expressed explicitly in terms of current and previous shocks. A linear model can be written as the output z_t from the linear filter

$$z_t = a_t + \psi_1 a_{t-1} + \psi_2 a_{t-2} + ...$$

$$= a_t + \sum_{j=1}^{\infty} \psi_j a_{t-j}$$

$$= \psi(B) a_t$$

whose input is a white noise, or a sequence of uncorrelated shocks a_t with mean 0 and common variance σ_a^2. Operating on both sides of the ARIMA model with the generalized autoregressive operator $\varphi(B)$, then

$$\varphi(B) z_t = \varphi(B) \psi(B) a_t$$

However, since

$$\varphi(B) z_t = \theta(B) a_t$$

it follows that

$$\varphi(B) \psi(B) = \theta(B)$$

Therefore, the φ weights may be obtained by equating coefficients of B in the expansion

$$\left(1 - \varphi_1 B - \ldots - \varphi_{P+d} B^{p+d}\right)\left(1 + \psi_1 B + \psi_2 B^2 + \ldots\right)$$

$$= \left(1 - \theta_1 B - \ldots - \theta_q B^q\right)$$

Thus, the φ_j weights of the ARIMA process can be determined recursively through the equations

$$\psi_j = \varphi_1 \psi_{j-1} + \varphi_2 \psi_{j-2} + \ldots + \varphi_{p+d} \psi_{j-p-d} - \theta_j \quad j > 0$$

with $\psi_0 = 1$, $\psi_j = 0$ for $j < 0$, and $\theta_j = 0$ for $j > q$. It is noted that for j greater than the larger of $p + d - 1$ and q, the ψ weights satisfy the homogenous difference equation defined by the generalized autoregressive operator, that is,

$$\varphi(B) \psi_j = \phi(B)(1 - B)^d \psi_j = 0$$

where B now operates on the subscript j. Thus, for sufficiently large j, the weights ψ_j are represented by a mixture of polynomials, damped exponential, and damped sinusoid in the argument j.

MOVING AVERAGES

This moving average is also referred to as smoothing technique. This method is suitable for forecasting data with no trend or seasonal pattern. The technique is a time series in a sequence of observations, which are ordered in time. Inherent in the collection of data taken over time is some form of random variation. The technique, when properly applied, reveals more clearly the underlying trend, seasonal and cyclic components of a rime series. The moving average is the best-known forecasting method. It simply takes a certain number of past periods and adds them together, then divides by the number of periods. Simple moving averages (MA) is effective and efficient approach, provided the time series is stationary in both mean and variance. The following formula is used in finding the moving average of order n, MA(n) for a period $t + 1$,

$$\text{MA}_{t+1} = \left[D_t + D_{t-1} + \ldots + D_{t-n+1} \right]/n$$

where D_i are past periods, and n is the number of observations used in the calculation. The forecast for time period $t + 1$ is the forecast for all future time periods. However, this forecast is revised only when new data becomes available.

The weighted moving average is another form of moving averages. It is very powerful and economical. They are widely used where repeated forecasts are required and use methods like sum-of-the-digits and trend adjustment methods. As an example, a weighted moving average is:

$$\text{Weighted MA}(3) = w_1.D_t + w_2.D_{t-1} + w_3.D_{t-2}$$

where the weights are any positive numbers such that: $w_1 + w_2 + w_3 = 1$. The average weights for this example are $w_1 = 3/(1 + 2 + 3) = 3/6$, $w_2 = 2/6$, and $w_3 = 1/6$.

ARTIFICIAL NEURAL NETWORKS

Artificial neural networks (ANNs) are used to generate a mapping between some input data and some required output. ANNs are model-free estimators, in that they do not rely on an assumed form from the underlying data. Rather, based on some observed data, they attempt to obtain an approximation of the underlying system that generated the observed data. They use nonlinear data-driven self-adaptive approach as opposed to the traditional model-based methods. They are powerful tools for modeling, especially when the underlying data relationships are known. ANNs can identify and learn the correlated patterns between the input data set and the corresponding target outputs. After training, the ANNs can be used to predict the outcome of new, unseen independent input data. ANNs imitate some aspects of the structure and learning of the human brain and can process problems involving nonlinear and complex data even when the data are imprecise and noisy. One of the appealing features of ANNs is that "learning by example" replaces "programming" in solving problems. This feature makes such computational models very useful in applications where one has little or incomplete understanding of the problem to be solved but where training data is readily available. ANNs are not intelligent, but they are good at recognizing patterns and making simple rules for complex problems. The neural network was historically inspired by the biological functioning of a human brain. Specifically, it attempts to mimic the fault tolerance and capacity to learn about biological neural systems by modeling the low-level structure of the brain. The brain is composed of a very large number of interconnected neurons. The neuron has a branching input structure (the dendrites), a cell body (the soma), and a branching output structure (the axon). The axon of one cell connects to the dendrites of another through a synapse, which is a junction between two nerve cells, consisting of a minute gap across which impulses pass by diffusion of a neurotransmitter.

When a neuron is activated, it fires an electrochemical signal along the axon. The signal crosses the synapses to the other neurons, which may in turn fire. A neuron fires only if the total signal received at the cell body from the dendrites exceeds a certain level (the firing threshold). The chance of firing, that is, the strength of the signal received by a neutron depends on the efficacy of the synapses. When creating a functional model of the biological neuron, there are three basic components of importance.

(i) **Synapses**: This is modeled as weights in ANNs. The strength of the connection between an input and a neuron is represented by the value of the weight. Unlike the synapses in the brain, the synapse weight of the artificial neuron lies within the range of positive and negative values. Positive weight values designate excitatory connections while negative values reflect inhibitory connections.

The next two components model the activity within the neuron cell:

(ii) **Linear combination**: This is an adder to sum up the input signal modified by their respective weights.

(iii) **Activation function**: This controls the amplitude of the output of the neuron. An acceptable range of output is usually between 0 and 1, or -1 and 1. Mathematically, a neuron k can be written as:

$$u_k = \sum_{j=1}^{m} w_j x_j$$

and

$$y_k = \varphi\left(u_k + b_k\right)$$

where $x_1, x_2, \ldots, x_m$ are the input signals; $w_1, w_2, \ldots, w_m$ are the respective synaptic weights of neuron k; u_k is the linear combination output due to the input signal; b_k is the bias; $\varphi(.)$ is the activation function and y_k is the output signal of the neuron. The use of bias b_k has the effect of applying an affine transformation to the output u_k:

$$v_k = u_k + b_k$$

where v_k is termed as the induced local field or activation field of neuron k. The linear combiner u_k is modified by b_k. A bias is similar in function to a threshold and is treated as a weight connected to a node that is always on. The weights determine where this hyperplane lies in the input space. Without a bias term, this separating hyperplane is constrained to pass through the origin of the space defined by the inputs. For some problems this is acceptable, but in many problems the hyperplane would produce increased performance away from the origin. If there are many inputs in a layer, they share the same input space and without bias they would all be constrained to pass through the origin. There are two fundamental different classes of network architecture:

FEEDFORWARD Anns

In this type of topology, the connections between the neurons in an ANN flow from input to output only. These ANNs can be further divided into either single-layer feedforward ANNs or multi-layer feedforward ANNs. The single-layer network is the simplest form of a layer network that has only one input layer that links directly to the output layer. With multi-layer feedforward ANNs, one or more hidden layers are present between the input and output layers by adding one or more hidden layers, the network can extract higher-order statistics from its input and model more complex nonlinear models.

FEEDBACK/RECURRENT Anns

In feedback or recurrent ANNs, there are connections from later layers back to earlier layers of neurons. There is at least one feedback loop in this type of network. Either the network's hidden neuron unit activation or the output values are fed back into the network as inputs. The internal states of the network allow this type of network to exhibit dynamic behavior when modeling the data's dependence on time or space. With one or more feedback links whose state varies with time, the network has adjustable weights. This results in the state of its neuron being dependent not only on current input signal but also on the previous states of the neuron. In other words, the network behavior is based on the current input and the results of previous processing inputs.

The learning process through which ANNs function can be categorized as supervised learning and unsupervised learning. For the purpose of clarity, the two classes of learning are explained here.

SUPERVISED LEARNING

This form of learning can be regarded as analogous to learning with a teacher, whereby the teacher has the knowledge of the environment. The knowledge is represented as input-output combinations. This environment is unknown to the neural network system. The teacher who has the knowledge of the environment will provide the neural network with desired responses for the training vectors. The network parameters are adjusted step by step under the combined influence of the training vector until the network emulates the teacher, producing the desired outputs for the corresponding inputs. At this stage, the network is presumed to be optimum in some statistical sense. In this way, the knowledge of the environment available to the teacher (ANN parameters frozen) is transferred to the neural network through training and stored in the form of fixed synaptic weights, representing long-term memory. When this condition is reached, the network is released from the teacher to deal with the environment by itself. With an adequate set of input-output examples, and enough time in which to do the training, a supervised learning system is usually able to approximate an unknown input-output mapping reasonably well.

UNSUPERVISED LEARNING

In this form of learning, the networks learn on their own as a kind of self-study. When a set of data is presented to the network, it will learn to recognize patterns in the data. To perform unsupervised learning, a competitive-learning rule may be applied, for example, by creating a neural network with two layers—an input layer and a competitive layer. The input layer will receive the available data. The competitive layer consists of neurons that compete with others for the opportunity to respond to features contained in the input data. The output of the network is not compared with the desired output. Instead, the output vector is compared with the weight vectors leading to the competitive layer. The neuron with the weight vectors most closely matching the input vector is the winning neuron. In other words, the network operates in accordance with a "winner-takes-all" strategy. Once the network architecture is decided and the data needed are collected, the next phase of the ANN methodology is the training of the ANN model. The training goal is to find the training parameters that result in the best performance, as judged by the ANN's performance with unfamiliar data. This measures how well the ANN will generalize. To find the optimum ANN configuration, an ideal approach is to divide the data into three independent sets: training, validation, and testing. The definitions of these terminologies are given as follows:

> **Training set**: A set of examples used to adjust or train the weights in the ANN to produce the desired outcome.
> **Validation set**: The validation error is used to stop the training. The validation error is monitored to determine the optimum point to stop training. Normally, the validation error will decrease during the initial phase of

training. However, when the ANN begins to overfit the data, the output error produced by the validation set will begin to rise. When the validation error increases for an appreciable number of iterations, thus indicating the trend is rising, the training is halted, and the weights that were generated at the minimum validation error are used in the ANN for the operation.

Testing set: To assess the performance of the ANN. As the real prediction accuracy will be generally worse than that for the holdout example, there is a need to evaluate the developed model with some real problem. In this research, the set of independent data used will be termed "evaluation sets." The "evaluation sets" used in this research represent two real problems, that is, predicting electricity consumption from two different data which were not used during the training process.

BIG DATA

For most organizations, big data is the reality of doing business. It is the proliferation of structured and unstructured data that flood organizations on a daily basis, and if managed well can provide powerful insights. Data is flowing and the volume is growing. With the massive generation of information from the advent of the Internet and the increasing digitalization of business, there is tremendous opportunity in the new amounts and types of data collected. But this data explosion has also dramatically increased the complexity of the enterprise data landscape, with multiple data lakes, data warehouses, operational applications, e-commerce, online interactions, and so on.

Big data technologies lack enterprise governance, holistic lifecycle management, and security concepts. These providers are just coming up the curve and trying to provide the level of enterprise governance and security that enterprise data warehouse and database providers have been delivering for their offerings for years. That means that organizations are stuck with limited tools for integrating systems and creating data pipelines. As a result, it takes a lot of effort to create a data pipeline across the enterprise. Some companies have tried to solve these issues by maintaining two sets of data: one for transactions and one for analysis. But this is not only costly and inefficient; it also leads to discrepancies because it's hard to keep them in sync. And discrepancies lead to inaccurate analytical outcomes, with the obvious negative impact on decision-making.

CHALLENGES OF BIG DATA

Meeting the needs of business and the fast pace of today's demands means that enterprise data needs to overcome the following three challenges:

1. **Governance**
   We face the lack of visibility, and ask: Who changed the data? What was changed? Who is accessing it?

2. **Data pipeline**

It is difficult to refine and enrich data across multiple systems. For example, this might involve improving the value of existing data by appending information, such as connecting sensor data with the asset ID and asset profile information, held in a different system.

3. **Data sharing**

Unfortunately, integration is manual, point-to-point, painful, and slow. Changing an integration point usually depends on the agility and flexibility of the IT line.

BIG DATA MANAGEMENT STRATEGY

The solution to better address these challenges would be a big data landscape and data operations management solution that enables agile data operations across the enterprise, and also enables data governance, pipelining, and sharing of all data in the connected landscape. The vision should be to provide the ability to understand, connect, and drive processes across the multiple data sources and endpoints with which the enterprise struggles today. By providing visibility into the landscape of data opportunities, as well as providing an easy way to connect data sources and create powerful data pipelines that hop across the landscape, businesses would be able to better achieve the data agility and business value that they seek.

ENTERPRISE RESOURCE PLANNING

Enterprise resource planning (ERP) is a process by which a company (often a manufacturer) manages and integrates the important aspects of its business. An ERP management information system integrates areas such as planning, purchasing, inventory, sales, marketing, finance, and human resources. With cloud computing now considered mainstream, more and more organizations are embracing cloud ERP (public and private) to drive enterprise-wide innovation to improve customer relationships while delivering operational efficiencies.

This trend is set to continue, with analysts predicting that cloud ERP adoption globally will continue to accelerate in all markets for the foreseeable future. Cloud's ascendancy in the enterprise is also being driven by its business relevance. Where once cloud was just another emerging technology, it has now become the key contributor and enabler of many of the current technology megatrends including hyper-connectivity, Internet of Things, big data, and social media.

Despite the continuing growth of adoption rates, most organizations still lack a coherent cloud application strategy on which to execute. Many are jumping headfirst into the cloud with little consideration for its broader business value, with the primary focus on introducing new business capability or only replacing applications reaching end of life. This approach has the potential to undermine the true business value, as opportunities for tangible savings can be realized only once a critical mass of applications and data have been moved to the cloud. As such, a few key fundamental elements need to be clearly defined:

- A robust cloud strategy
- A supporting application and information strategy
- Establishment of a cloud-ready organization—people and process

These elements provide consistency in understanding the business drivers for cloud adoption, the architectural principles and framework relevant to the cloud, capabilities required both internal and external to the organization, approach to commission new and existing applications, and how risks associated with migration will be managed.

BUILDING THE BIG DATA WAREHOUSE

The enterprise data warehouse (EDW) architecture has long been a key technology asset for fast analytics on cleansed, curated, and structured business data. It is a critical technology foundation of many enterprises. However, it is straining to deliver value in the era of exploding data volumes and increased demand for analytics and data across the organization. Not only are the sources of structured, traditional data increasing in volume—such as transactional, operational, and financial information—but organizations are also embracing the age of big data, dealing with new types of data that the enterprise data warehouse was not built to handle. This includes unstructured information, like weblog and machine log information, audio, video, and social media interactions; high-speed information, like sensor data in Internet-of-Things scenarios; and third-party information, like weather, public databases, or brokered information. All of this data is being introduced to the enterprise at unprecedented volume and speed. This big data is stored in systems uniquely capable of handling them, such as Hadoop-based data lakes or cloud object storage.

The business potential inherent in all of this data is demanding to be tapped, not only to improve the efficiency and quality of existing goods and services but also to create new offerings or business models that can accelerate an organization ahead of the competition. However, in order to achieve this, enterprises need a way to interconnect big data with enterprise data. They also need a way to provide the analytical responsiveness, security, and ease of use that are associated with the enterprise data warehouse and its applications. With a big data warehouse approach, companies are looking to:

- Leverage existing investments made in technology, processes, and people. While organizations recognize the limitations of the EDW in the face of new data demands, they also recognize the value of the data already being managed effectively by the EDW and want to leverage this as part of their new enterprise architecture. There is not only the financial investment in the technology itself to consider but also the value of the existing processes that are well understood by the enterprise's employees, as well as partners and vendors. While change and new investment are inevitable, leveraging the EDW to its maximum potential makes clear sense from a financial and change management perspective.

- Innovate for the future, leveraging new and faster sources of data. Enterprises are increasingly embracing big data, especially as big data solutions become easier to use, more secure, and better integrated to traditional enterprise systems. These new, faster data sources represent greater opportunities for improving existing products and services, as well as an opportunity to capture new and emerging markets. Enterprises across industries increasingly see competitive threats from disruptive new entrants that are effectively using new data architectures built for big data from the start.

- Make faster, more responsive, and even proactive decisions. End customer expectations for service speed, corporate responsiveness, and information-sharing from the companies that serve them are increasing quickly. As a result, enterprises are looking to deliver on expectations by accelerating their own speed of data collection, processing, and analysis, aiming to spread "right time" data and decision-making as broadly across their organizations as they can. Forward-looking organizations have future goals not only for responsiveness to rapidly changing market or customer conditions but also for achieving predictive analytics that can help them to address issues before they escalate, or spot trending opportunities in their earliest stages.

- Empower more managers and decision-makers with analytics. "Self-service" analytics has been a goal of enterprises for some time now, and the intensity of pressure for achieving this goal has only increased. Organizations are realizing that limiting analytics and decision-making to an elite few requires too much time and can squander opportunities or exacerbate emerging challenges. By more broadly distributing information and analysis to a wider range of managers and decision-makers, including partners and key vendors, the organization becomes more responsive and agile.

- Ensure enterprise-class security and data governance, even as data volumes grow and analytical end users proliferate. While big data solutions have long offered a scalability and data diversity advantage that EDWs could not match, there have also been longstanding concerns about data security and data governance for these emerging technologies. As the technologies mature and can increasingly meet the security needs of the most demanding organizations, there is increased willingness to let big data projects out of the "lab" and interconnect them with the broader enterprise data architecture and its larger group of end users.

BIG DATA ANALYTICS

Big data analytics refers to the strategy of analyzing large volumes of data, or big data. This big data is gathered from a wide variety of sources, including social networks, videos, digital images, sensors, and sales transaction records. The aim of analyzing all this data is to uncover patterns and connections that might otherwise be invisible and that might provide valuable insights about the users who created it. Through this insight, businesses may be able to gain an edge over their rivals and make superior business decisions.

Big data analytics allows data scientists and various other users to evaluate large volumes of transaction data and other data sources that traditional business systems would be unable to tackle. Traditional systems may fall short because they're unable to analyze as many data sources. Sophisticated software programs are used for big data analytics, but the unstructured data used in big data analytics may not be well suited to conventional data warehouses. Big data's high processing requirements may also make traditional data warehousing a poor fit. As a result, newer, bigger data analytics environments and technologies have emerged, including Hadoop, MapReduce, and NoSQL databases. These technologies make up an open-source software framework that's used to process huge data sets over clustered systems.

OPEN-SOURCE BIG DATA ANALYTICS

Open-source big data analytics refers to the use of open-source software and tools for analyzing huge quantities of data in order to gather relevant and actionable information that an organization can use in order to further its business goals. Open-source big data analytics makes use of open-source software and tools in order to execute big data analytics by either using an entire software platform or various open-source tools for different tasks in the process of data analytics. Many big data analytics tools make use of open source, including robust database systems such as the open-source MongoDB, a sophisticated and scalable NoSQL database very suited for big data applications, as well as others. Open-source big data analytics services encompass:

- Data collection system
- Control center for administering and monitoring clusters
- Machine learning and data mining library
- Application coordination service
- Compute engine
- Execution framework

DATA VISUALIZATION

Data visualization is the presentation of data in a pictorial or graphical format. It enables decision-makers to see analytics presented visually, so they can grasp difficult concepts or identify new patterns. With interactive visualization, technology can be used to drill down into charts and graphs for more detail, interactively changing what data is seen and how it's processed. With big data there's potential for great opportunity, but many retail banks are challenged when it comes to finding value in their big data investment. For example, how can they use big data to improve customer relationships? How—and to what extent—should they invest in big data?

Because of the way the human brain processes information, using charts or graphs to visualize large amounts of complex data is easier than poring over spreadsheets or reports. Data visualization is a quick, easy way to convey concepts in a universal manner—and you can experiment with different scenarios by making slight adjustments. Data visualization performs the following functions:

- Identify areas that need attention or improvement.
- Clarify which factors influence customer behavior.
- Help understand which products to place where.
- Predict sales volumes.

IMPORTANCE OF DATA VISUALIZATION

Data visualization has become the de facto standard for modern business intelligence (BI). Data visualization tools have been important in democratizing data and analytics and making data-driven insights available to workers throughout an organization. They are typically easier to operate than traditional statistical analysis software or earlier versions of BI software. This has led to a rise in lines of business implementing data visualization tools on their own, without support from IT.

Data visualization software also plays an important role in big data and advanced analytics projects. As businesses accumulated massive troves of data during the early years of the big data trend, they needed a way to quickly and easily get an overview of their data. Visualization is central to advanced analytics for similar reasons. When a data scientist is writing advanced predictive analytics or machine learning algorithms, it becomes important to visualize the outputs to monitor results and ensure that models are performing as intended. This is because visualizations of complex algorithms are generally easier to interpret than numerical outputs.

USAGE OF DATA VISUALIZATION

Regardless of industry or size, all types of businesses are using data visualization to help make sense of their data. This is achieved as follows:

1. **Comprehend Information Quickly**

   By using graphical representations of business information, businesses are able to see large amounts of data in clear, cohesive ways—and draw conclusions from that information. And since it's significantly faster to analyze information in graphical format (as opposed to analyzing information in spreadsheets), businesses can address problems or answer questions in a more timely manner.

2. **Pinpoint Emerging Trends**
   Using data visualization to discover trends—both in the business and in the market—can give businesses an edge over the competition, and ultimately affect the bottom line. It's easy to spot outliers that affect product quality or customer churn and address issues before they become bigger problems.
3. **Identify Relationships and Patterns**
   Even extensive amounts of complicated data start to make sense when presented graphically; businesses can recognize parameters that are highly correlated. Some of the correlations will be obvious, but others won't. Identifying those relationships helps organizations focus on areas most likely to influence their most important goals.
4. **Communication**
   Once a business has uncovered new insights from visual analytics, the next step is to communicate those insights to others. Using charts, graphs or other visually impactful representations of data is important in this step because it's engaging and gets the message across quickly.

CONCLUSION

Over time, businesses have gathered a large volume of unorganized data that has the potential to yield valuable insights. However, this data is useless without proper analysis. Modeling techniques can find a relationship between different variables by uncovering patterns that were previously unnoticed. For example, analysis of data from point of sales systems and purchase accounts may highlight market patterns like increase in demand on certain days of the week or at certain times of the year. Optimal stock and personnel can be maintained before a spike in demand arises by acknowledging these insights. Modeling is not only great for lending empirical support to management decisions but also for identifying errors in judgment. Modeling can provide quantitative support for decisions and prevent mistakes due to manager's intuitions.

REFERENCES

Badiru, Adedeji B., Oye Ibidapo-Obe, and Babs J. Ayeni (2019), *Manufacturing and Enterprise: An Integrated Systems Approach*, Taylor & Francis Group/CRC Press, Boca Raton, FL.

Box, G., Jenkins, G., and Reinsel, G. (2008). *Time Series Analysis: Forecasting and Control*. 4th Edition. John Wiley Publishers, New York, pp. 100–105.

9 Case Study of Systems Drivers in Aerospace Product Development

INTRODUCTION

Product development in a systems environment is of high interest to researchers and practitioners. Badiru (2012) presents the application of the DEJI Systems Model® to the complexity of product development in the aerospace industry. Most of this chapter is an adaptation of the case study presented by Jaifer et al. (2021), in which the drivers of efforts and time in aerospace product development projects are extensively analyzed. These two references (Badiru, 2012 and Jaifer et al., 2021) align perfectly for the case-study theme presented in this chapter. Based on the case study presented by Jaifer et al. (2021), product development projects, especially in the aerospace industry, suffer from significant cost and schedule overruns. Readers interested in the complete details of the case study should refer to Jaifer et al. (2021) and all the references, therein. In the case study, Jaifer et al. (2021) assert that many researchers and scientists focused first on processes and tools to shorten the time and reduce the cost of new product development projects. The research results did not resolve the perennial challenge. Thus, the studies and publications continue. This chapter adds to that body of effort. A large number of research papers have been published regarding effort and time estimation in new product development. Most of them are based on the use of some specific drivers for the estimation objective. Thus far, there are no rules or frameworks about which drivers are relevant to use and which are not. The case study by Jaifer et al. (2021) presents a new and original framework that characterizes and organizes effort and time drivers in aerospace product development. The case study complements the structured framework embodied in DEJI Systems Model. The aim of Jaifer et al. (2021) is to identify and support the understanding of most relevant drivers for aerospace product development effort and time. The framework was developed based on an extensive literature review and a postmortem analysis of cost and time overruns of a significant product development program in the aerospace industry. The final list of the frameworks' drivers was validated and refined using a survey. Results suggested that besides risks and uncertainty, technologies maturity, degree of change in design, ambiguity of requirements, functional decompositions, severity of standards, process overlapping and variety of key stakeholders drive effort and time as complexity drivers while processes maturity, experience with technology, risk management, change management, level of trust in suppliers, and team skills drive effort and time as proficiency drivers.

DOI: 10.1201/9781003175797-9

Failure to meet planned cost and time is a common issue in new product development Among other reasons, accuracy of estimates comes as one of the most popular. If the cost and time are estimated within a reasonable degree of accuracy, the ability to schedule, forecast, and conduct trade-offs, among others, will become much easier (Jaifer et al., 2021). Otherwise, poor estimations could even lead, in some cases, to severe difficulties due to wrong financial decisions made about engagement and partners' relationship management. Thus, one of the most important questions that confront the program manager in the planning phase of new product development is "how to get the right estimates of cost and time?" Different approaches, methods, and models were developed for this purpose.

Despite the rich literature about these models, specifically parametric ones, it is still unclear how researchers chose the right parameters or drivers for their models. In other words, how can the right drivers for estimations models be identified? Is it sufficient to use only some critical drivers? Is there a broadly accepted list of potential drivers to select from? What are the different classes of drivers? As a whole, is there any framework that structures potential drivers and groups them by categories, by classes, or by dimensions and then highlights how they are linked to effort or time? From the authors' point of view, there is a lack of importance given to drivers' selection as core elements in estimation models. In fact, the literature review shows that almost all research focused on parametric modeling used a few variables or drivers to predict cost and time. The choice and relevancy of these variables were, for the most part, not justified from a scientific or research point of view. In other words, the choice was not based on a broadly recognized framework or selection criteria. Most of the models use previous experience and/ or internal experts' judgment to decide which drivers or parameters will determine effort and time (Jaifer et al., 2021). The process of estimating effort and time in new product development would require a new framework for potential drivers of cost and time. Such a framework should group and structure potential drivers and eventually identify their relevance for more accurate estimation results. Since the time in person-hours, also known as effort make up the bulk of the cost for most product development projects, the effort can provide a good picture of project cost (Jaifer et al., 2021). For the rest of this paper, the term "effort" is used as a substitute for the term "cost" in product development.

This paper aims to develop a new framework for the potential drivers of effort and time in new product development. It notably focuses on the aerospace industry. For this purpose, the research effort started with an extensive literature review exploring drivers used by different estimation models. The interest was also focused on the key concepts involved in aerospace product development performance, which in turn are related to effort and time drivers. This concerns especially concepts like complexity, uncertainty, and proficiency. As a result, the framework is structured regarding these concepts. The framework was developed as a starting point to help in bringing consensus within the planning team about effort and time drivers in the aerospace industry. It will help both practitioners and researchers to choose the most relevant drivers for system parameter estimation purposes.

CASE STUDY BACKGROUND

To develop a new framework for effort and time drivers as mentioned before, the research was based both on theory and practice. First, an integrative literature review (Jaifer et al., 2021) was performed with the aim to gather potential drivers of effort and time estimations. Thereby, underlying concepts related to effort and time in aerospace product development were also covered by the literature review. The objective was not to go in depth with these concepts but rather to use them to identify more drivers and, especially, to define potential classes and categories of drivers. This helped in developing the building blocks and structure the framework through concepts' integration. In parallel to the literature review, an industrial case study was performed with the objective of bringing practical insight into the methodology. This case study is the first part of the fieldwork and concerns a postmortem analysis of a five-billion-dollar new product development program in the aerospace industry. The program is composed of ten work packages (WP). Each of the WPs has *its* own organization and focuses on the development of a part or set of parts of the final product. The case study allowed for the identification of the main elements contributing to cost and schedule overruns which in turn led to the identification of effort and time drivers. The objective was to identify the most relevant drivers through data analysis, semi-structured interviews, and work*shops*. Semi-structured interviews concerned mainly people involved in planning management, suppliers' management, change management, cost management, test, and validation. The selection of participants was based on internal experts' recommendations and elements that evolved from data analysis. Workshops involved employees within the same department or work package but with different backgrounds and over several levels of authority. In total, 11 workshops and more than 50 semi-structured interviews were conducted. Results from the case study helped complete the list of drivers found in the literature from practical insights. The results from the literature search and case study were brought together to develop the preliminary framework with a detailed list of potential drivers of effort and time.

The second part of the fieldwork aim was to evaluate the relevance or criticality of elements resulting from the literature review and case study through a survey involving 29 experts from the aerospace industry. The selection criteria for participants were mainly based on their involvement in cost and time planning of new product development. The main objective of the survey was to refine the framework and highlight critical elements through a relevancy evaluation.

PRODUCT DEVELOPMENT IN THE AEROSPACE INDUSTRY

New product development projects have vital importance in the aerospace industry. In fact, products are a means by which companies achieve their objectives in terms of revenues (Jaifer et al., 2021). Cooper's research shows that in common industry one-third of revenues should come from new products. In dynamic industries, even 100% of revenues can stem from new products. However, the

NPD project's success rate in terms of cost and time remains low. The study realized by G. Bounds (Jaifer et al., 2021) showed that only 26% of new development projects respect cost and schedule. The U.S. Government endeavor to predict cost overrun for aerospace and defense new product development resulted in a new law called the Nunn-McCurdy Act (Jaifer et al., 2021). The main objective of this law is to set up some rules and requirements about cost overrun and its metrics in large-scale aerospace and defense product development projects (Jaifer et al., 2021). This shows clearly that cost and schedule overruns are the main issues in aerospace product development. Overruns are mainly due to the extremely constrained environment and the complexity of product development in this industry. The following are examples of most influencing constraints and challenges facing aerospace companies for developing their products.

The product development environment in the aerospace industry is very particular, especially with the implication of a large number of stakeholders, including in most cases the government. For these reasons and regarding the nature of aerospace products, regulation and laws are very strict. Despite the diversity of stakeholders, the single customer is very common in the aerospace business model. In fact, aerospace products are in general developed for one or very specific customers. This restricted market constraint prevents aerospace companies from exploring the lever of economies of scale which should allow for the amortization of R&D cost and lead to low-cost product unit. Unfortunately, in the aerospace industry, this is not possible and product development nonrecurring cost has a great impact on cost unit, and then on sale price, which explains the high market risk.

The competitive dynamics of the aerospace sector and the evolution of customers' requirements force most companies to continuously bring significant improvements to the performance and capabilities of their products, which leads to a fast evolution of technologies. Thus, aerospace companies seek to be at the forefront of technological innovation and develop products with next-generation technologies that they should mature during the product development cycle (Jaifer et al., 2021). This creates technological complexity during the product development process. Product multi-disciplinarity and issues of technology integration intensify this complexity.

Besides technological issues, product development in the aerospace industry runs through a complex organizational system of processes, tools, and resources involving different disciplines with many interfaces and must naturally develop interrelated subsystem (Jaifer et al., 2021). This is why interactions levels are very high in the aerospace product development organization. Multi-disciplinarity increases the complexity of the product development process and organization (Jaifer et al., 2021). Some practices like iterative development and process overlapping create more interactions during product design and bring another dimension of complexity to the product development organization (Jaifer et al., 2021). From the time perspective, products and systems in the aerospace industry tend to get long duration to conceive, develop, and build. The lifecycle differs from one system to another, but it's not uncommon for 5 years to even a complete decade between concept and product launch. The challenge is how to get a clear and correct understanding of the evolution of the environment in which these products will operate. The change management process plays a key role in product development to bring the required adjustments, but the efficiency of this process is a paramount factor to secure these adjustments, without losing control over product development cost.

The level and diversity of uncertainty sources are the most important issues that are intrinsically related to product development in the aerospace industry. In fact, development risk and uncertainty are so extreme that many companies cannot build a product without customers' co-investment and/or government risk-sharing contract. Uncertainty is caused by a lack of sufficient information (Jaifer et al., 2021). From the product development perspective, it's related to the lack of information about requirements: technologies, environment, and the product in general.

In the aerospace sector, uncertainty sources are highly diversified by the fast evolution of technologies and customers' requirements, the variety of stakeholder's expectations combined to the time frame of product development project have the greatest impact on uncertainty and risk level. Many research studies confirmed the effect of uncertainty on cost and schedule in product development (Jaifer et al., 2021). It was reported that uncertainty is the main motivation for engineering change which in turn should result in cost and schedule overruns. According to T. M. Williams (Jaifer et al., 2021), increased uncertainty would contribute to product development complexity and then increase the chance of cost and schedule overrun.

Uncertainty and complexity remain some of the key concepts in aerospace product development, considering their effects on cost and schedule. Therefore, besides performance factors that have an obvious effect on effort and time, the concepts of complexity and uncertainty should be considered as the key cornerstones of any framework structuring effort and time drivers in aerospace product development.

EFFORT AND TIME DRIVERS IN THE LITERATURE

An extensive literature review about drivers of effort (or cost and time) in new product development was the starting point of this research study. It helped identify, directly and indirectly, a non-exhaustive list covering almost all new product development effort and time drivers used in the literature. The literature search methodology comprises the search terms, resources, and the search process. The search terms are naturally derived from the research objective, which is the identification of relevant effort and time drivers for potential use in estimation models. The use of the Boolean OR allows the incorporation of alternative spellings and synonyms while the use of the Boolean AND allows linking the terms.

A further literature review of key concepts related to effort and time in product development helped not only in completing the list but also in categorizing drivers and in enlightening relationships among them. The search process was conducted in three steps. In the first step, a search in each database was performed using the search terms and provided a set of potential papers about effort and time drivers in estimation models. The lists of references in relevant papers were examined to extract more potential papers and ensure that the search covered the maximum number of studies related to effort and time drivers The second step concerned the examination and selection of relevant papers based on two criteria:

(1) studies about efforts and/or time estimation models using one or more drivers.
(2) studies concerning product development project preferably in the aerospace or technical sector.

Indeed, as there is a perceived lack of framework that defines the structure and categories of effort and time drivers, the aim of this step was to identify drivers' categories and then conduct an in-depth search about these categories or related concepts, while aiming for identifying more relevant effort and time drivers. The aim was also to define an initial structure for the effort and time driver's framework. Reviewed papers about these concepts should have an effort and time perspective and hopefully should help identify more drivers. The overall search process resulted in 127 papers, of which 36 are related to estimation models and help directly in identifying drivers of effort and time. The others helped either with structuring the framework or indirectly identifying more drivers.

ESTIMATION MODELS FOR EFFORT AND TIME DRIVERS

Notwithstanding the abundance of research works about effort and time estimation in new product development, none of the reviewed papers tried to scientifically define relevant drivers to be used, or at least formalize a process for selecting these drivers. In most cases, this is determined by internal consensus, experts' judgment, or authors' judgment. M. F. Jacome and V. Lapinskii (Jaifer et al., 2021) are among those authors who are interested early in developing a cost estimation model in product development Their model was focused on electronic products design and took into account three main drivers: size, complexity, and productivity. The first driver represents the size of the product as reported by the number of transistors, the second driver considers tasks' difficulties, and the third driver represents the rate (effort by transistor) at which the task progresses. Despite the use of generic drivers for this model, their valuation was adjusted to electronic products design.

H. Bashir and V. Thomson (Jaifer et al., 2021) are among the authors who are interested in cost estimation modeling for design activities. They proposed different models over 5 years of research effort. Their first model for time estimation is a generic one and could be applicable for a wide range of design projects with different sizes and over different sectors. Authors reported that the choice of drivers is made from hundreds of influencing factors on different aspects of process design. After reviewing published research, the six following drivers were identified as more relevant and as major contributors to design effort variation: product complexity, technical difficulty (severity of requirement and use of new technologies), skills and experience of team members, the use of design-assisted tools, and finally the use of formal processes. Regarding many considerations like the characteristics of historical projects, statistical constraints, and some theories concerning design projects, the choice of drivers was limited to the complexity of the product, as presented by the functional structure and the severity of requirements.

The second effort estimation model of H. A. Bashir and V. Thomson is destined especially for GE Hydro projects (Jaifer et al., 2021). For this model, the choice of drivers was based on a review of published researches and through consulting with design managers from GE Hydro in a brainstorming session. The list of drivers was limited to the following: product functionality structure, technical difficulty versus team expertise, type of drawing, and involvement of design partners. Bashir & Thomson reported that product functionality as an indicator of product complexity accounted for about 80% of the variation in project effort.

P. Duverlie and J. M. Castelaio (Jaifer et al., 2021) used product features or product physical parameters as drivers of cost in the design process. They compared the parametric modeling with base case estimation in mechanical design. D. Xu and H. S. Yan (Jaifer et al., 2021) developed a new model for estimation of product design time using neural networks and fuzzy logic, which offers the possibility of integrating more drivers than parametric modeling. Thus, based on different models that describe the product development cycle time, more than 20 drivers were identified and classified into the following seven categories: product characteristics, design process, design conditions, design team, project complexity, information process capability, and motivation. The authors recognized the difficulty in evaluating and measuring some drivers at an earlier stage of the design process, thereby, they proposed methods for quantitative drivers' evaluation from qualitative information using the house of quality and the fuzzy logic approaches.

L. Qian and D. Ben-Arieh (Jaifer et al., 2021) proposed a parametric cost estimation model, which combines the activity-based costing (ABC) approach with the feature-based approach. The latter used basically physical characteristics of the product such as volume or mass as cost-related features or cost drivers in parametric modeling. Thus, the model identified each activity in the product development process and then defined one or more cost drivers for each activity. All drivers were in general product or tools related features.

S. R. Meier (Jaifer et al., 2021) conducted a research study to ascertain the main causes of cost and schedule growth in large-scale programs of the Department of Defense. The seven following causes were reported by the author as main contributors to cost and schedule growth: immature technology, lack of corporate technology roadmap, requirement instability, ineffective acquisition strategy, inadequate system engineering, workforce issues, and unrealistic program baseline. This study highlighted some relevant cost and time drivers specific to aerospace and defense product development. It showed the importance of technology maturity and technology management and their effect on cost and schedule. It also mentioned the role of requirements management and acquisitions in cost and time performance. Thus, it's a very inspiring study for the identification of cost and time drivers in aerospace product development projects.

A. Salam et al. (Jaifer et al., 2021) developed a design effort estimation model with a focus on the aerospace industry. The model represents a modified version of H. A. Bashir and V. Thomson's model (Jaifer et al., 2021). The authors kept the structure or the cost estimation relationship (CER)of the aforementioned model with the integration of new effort drivers. In fact, the product complexity was omitted as the model was applied to a specific type of product and then this driver was no more a variable. Based on extensive interviews and discussions with managers, designers, and project engineers in the aerospace industry, the authors defined four drivers for the effort estimation model: type of design, the degree of change, concurrency, and experience of departmental personnel. The model was validated through a case study in the aerospace industry. A. Salam and N. Bhuiyan (Jaifer et al., 2021) conducted a new research study in effort estimation for the aerospace industry. They developed and compared different CER techniques for effort estimation, mostly linear and nonlinear regression techniques. They used the same drivers as the previous research work namely the type of design, the degree of change, concurrency, and the experience of departmental personnel. A. Sharma and D. S. Kushwaha (Jaifer et al., 2021) based their research on the estimation of development and testing effort on the complexity of requirements. They judged that the complexity of requirements has a direct bearing on the required effort in product development. They suggested an overall process for the computation of a new index named the improved requirement based complexity (IRBC). The computation of the IRBC index uses many parameters like the functional requirement complexity, the nonfunctional requirement complexity, the design constraints, the interface complexity, and features' complexity. Therefore, they combined the IRBC index with the productivity and the technical complexity factor to estimate the required effort. The technical complexity factor is evaluated according to some parameters like data communication, processing complexity, configuration, transactions rules, multiples site, and easiness of operations. The Sharma-and-Kushwaha proposal (Jaifer et al., 2021) is mainly based on the complexity concept for effort estimation. P. Chwastyk and M. Kolosowski (Jaifer et al., 2021) proposed a one-driver model for cost estimation in new product development The model is based on simple linear regression using a physical dimension of the product as the dependent variable. The authors illustrated the results of applying the model through the case of valve development. Thus, the valve diameter was used as the independent variable.

M. T. Adoko et al. (Jaifer et al., 2021) developed a cost overrun predictive model for complex system development like aerospace and defense systems. The aim of their research effort was to help foster a better understanding of factors that lead to cost and schedule overruns, which should highlight cost and schedule drivers in aerospace and defense new product development. The main factors judged to be strongly associated to cost and schedule overruns were technologies readiness level or technologies maturity, system engineering process, performance, reliability, and risk level. Besides their use in cost overrun prediction, these factors should also give a clear idea about some drivers or categories of drivers in aerospace product development effort and time estimation.

R. Vargas (Jaifer et al., 2021) used neural networks and analogous estimation to determine the project budget. Variables used in his model were limited to the level of project's complexity, location, baseline duration, type of contract, and number of relevant stockholders. The used data covered about 500 projects in different sectors.

CATEGORIZATION OF EFFORTS AND TIME DRIVERS

This first step of literature review allowed for the examination of 36 studies about effort/cost estimation models using from 1 to 15 drivers, for a total of 119 drivers. The primary aim of this step is the inventory and categorization of drivers as well as the identification of key concepts. Thereby, a semantic analysis of drivers' terminologies and keywords was conducted using synonyms, hyponyms, and hypernyms. This allowed connecting drivers with their surrounding concepts while facilitating the identification of concepts and categories. The categorization respected also the three-level coding guidelines (open coding, axial coding, and selective coding), as reported by A. L. Strauss and J. Corbin (Jaifer et al., 2021). Foremost, all drivers related to product characteristics or features like dimension, weight, etc., were renamed "product characteristics and features"; this is to avoid getting an unlimited number of potential drivers of this nature. Likewise, drivers that reflect technical specifications related to the nature of project or sector activities were renamed "technical specifications." To draw up the final list of drivers, redundant ones were eliminated, providing a total of 58 different drivers. Therefore, terminology mapping and analysis of these drivers' keywords helped clustering them by categories. These categories, in turn, helped identifying keys concepts related to effort and time.

From the list of 58 drivers, 10 contain the terms "complexity" or "difficulty." These drivers were then classified in the "complexity" category. Despite that only two drivers contain the term "size," seven others are related to the measure of project size like number of requirements and "number of pieces." These nine drivers were classified in the "size" category. Five of the drivers that represent risks and uncertainty were grouped in the "risk/uncertainty" category. A significant number of drivers contain terms like "performance," "productivity," "capability," "experience," and "expertise." This category encloses 19 drivers that are related to the performance, capability, and ability of external and internal resources (human, tools, processes, and suppliers) to execute projects' activities. This category was initially named performance/capability. Finally, through a careful examination of the 15 remaining drivers, taking into consideration their meanings as intended by the authors, we concluded that they represent some product and project conditions and attributes.

The main concepts that could be extracted from this preliminary classification are related to the "hyponym" under which the classification is done. This concerns complexity, size, risk/uncertainty, and performance/capability. However, project conditions and attributes drivers could not be associated with a specific concept related to their terminologies. According to the analysis of their causes/effects' relationship with effort and time and taking into consideration the meaning intended by the authors, these drivers, if not managed appropriately, will increase the effort and time required to execute the project activities through the increase of complexity. Thus, these drivers are, in fact, related to the concept of complexity. The literature review about complexity should bring about most of these drivers as complexity drivers. The category "performance/capability" was renamed "proficiency," as this is the closest concept that regroups performance, capability, productivity, and skills. The business dictionary defines proficiency as "[m]astery of a specific behavior or skill demonstrated by consistently superior performance, measured against established or popular standards"

As key concepts related to effort and time were identified, a second iteration of literature review concerning these concepts was undertaken with the aim to extract more drivers of effort and time.

COMPLEXITY OF EFFORT AND TIME DRIVERS

As reported earlier, complexity is a key concept in product development, especially in the aerospace sector. A large number of papers were published about this concept, demonstrating its evident importance. Many of these researches focused on complexity and uncertainty effects on project cost and schedule. Unfortunately, there is a lack of consensus about a standard definition of complexity. According to S. Schlindwein and R. Ison (Jaifer et al., 2021), research orientations about complexity are mainly divided into scientific approaches: descriptive and perceived complexity. The first approach considers complexity as a property of the system and tries to evaluate it through some factors. The latter considers complexity as subjective since it is improperly understood through the perception of an observer. This study notably focuses on the first approach.

L. A. Vidal and F. Marle (Jaifer et al., 2021) reported that complexity appears to be one of the main reasons for the unpredictability of projects. M. V. Tatikonda and S. Rosenthal (Jaifer et al., 2021) claim that complexity in NPD should result in unmet specifications and budget overrun. From T. Williams' point of view (Jaifer et al., 2021), the underestimation of project complexity is one of the main reasons for project failure in terms of cost overruns and schedule delay. H. ParsonsHann and K. Liu (Jaifer et al., 2021) reported that the complexity of requirements could contribute to project failures. K. Caniato and A. Grobler (Jaifer et al., 2021) illustrated the moderating effects of product complexity on new product development performance. For these reasons, we believe the complexity, through some factors, drives the effort and time in product development. Thus, for the purpose of our research study, we reviewed papers that focus on product development complexity with the hope to extract potential drivers of effort and time. We also highlighted the different classes or categories of complexity factors, as determined by the literature. This helped to classify drivers and facilitate the structuring of our framework.

The 1996 research study of D. Baccarini (reported by Jaifer et al., 2021) was among the first to review the concept of project complexity and objectively define complexity factors far from the subjective connotation of the meaning of complexity as defined in 1993 by T. M. Wozniak (see Jaifer et al., 2021). According to D. Baccarini (Jaifer et al., 2021), different types of complexity factors are most commonly recognized in project management. That is organizational complexity factors and technological complexity factors. H. A. Bashir and V. Thomson (Jaifer et al., 2021) were also among the first authors who investigated the complexity factors related to design activities. Their objective was to develop metrics to measure complexity. They reported that the functional structure of the product should be the most relevant index of design complexity. According to their research study, functional structure is the main factor of complexity in product design projects. The effort of J. D. Summers and J. J. Shah (Jaifer et al., 2021) to unify complexity measurement in design activities is noteworthy. The authors classified complexity factors or complexity measures into three categories: solvability, size, and coupling. They suggested that these measures should capture different aspects of product design and then have to be kept independent. The solvability reflects how the designed product may be predicted to satisfy the design problem.

F. Ameri et al. (Jaifer et al., 2021) investigated methods for measuring complexity in engineering design based on the valuation of size and coupling complexity. They proposed a coupling complexity measure to evaluate the decomposability of the graph-based representation of the design product. This means that the product hierarchical structure is a main factor of complexity. S. A. Sheard and A. Mostashari (Jaifer et al., 2021) proposed a typology of complexity in engineering systems consisting of three classes and six main subtypes. The typology includes three types of structural complexity (size, connectivity, and architecture), two types of dynamic complexity (short term and long term), and finally the socioeconomic complexity.

S. Tamaskar et al. (Jaifer et al., 2021) proposed a method for measuring the complexity of product design with a focus on aerospace systems. They determined seven aspects or classes of complexity factors to structure the framework. They judged that levels of abstraction, type of representation, size, heterogeneity of components and interactions, network topology factors such as coupling, modes of operation, and off-design interactions to be the most relevant factors for measuring complexity in aerospace product development. This framework combines structural and functional elements and addresses cross-domain interactions. Later, S. Tamaskar et al. (Jaifer et al., 2021) improved the measurement method by proposing a framework for measuring the complexity of the aerospace system. Their framework focused on measures that capture size, coupling, and modularity aspects. W. ElMaraghy et al. (Jaifer et al., 2021) reviewed the breadth of the complexity of the design product/process, manufacturing, and business environment. They identified related complexity factors or drivers for each category. As far as product development is concerned, product design and business environment complexity should play a key role. The number of parts, multidisciplinary, manufacturability, size/geometry, and variety were defined as complexity factors in product design, while global competition, market turbulence, foresight, and supply chain dynamics were defined as business environment complexity factors.

Many researchers went in depth in examining project or product development complexity. In fact, they developed project complexity frameworks by exploring complexity factors or drivers and eventually classified them by categories. L. A. Vidal and F. Marle (Jaifer et al., 2021) studied the project complexity with the purpose of identifying and modeling complexity within the field of project management in order to plan and manage better under conditions of W1certainty. Concretely, they defined a new framework for project complexity by using the D. Baccarini (Jaifer et al., 2021) traditional dichotomy classification of complexity factors (technological and organizational) and then combined it with their subclassification of four classes: project size, variety, interdependencies within project system, and context-dependent. A broad literature structured according to the four aspects of system thinking allowed for the identification of complexity factors. The framework was then built by splitting those complexity factors according to the classification. Authors claim that with this framework, the project complexity appears as multiple aspects or multiple criteria characteristics of the project. L. A. Vidal et al. (Jaifer et al., 2021) worked on a framework to propose a new method for project complexity measurement using the AHP process. In terms of complexity factors, they conducted a broad survey using a five-level Likert scale to judge the relevancy of each factor.

BADIRU'S PROJECT COMPLEXITY MEASURE

Related to the above exposition, it is recalled that as far back as 1984, Badiru (1988) developed a quantitative measure of the complexity of project networks. That work is described here. The performance of a scheduling heuristic will be greatly influenced by the complexity of the project network. The more activities there are in the network and the more resource types are involved, the more complex the scheduling effort. Numerous analytical experiments have revealed the lack of consistency in heuristic performances. Some heuristics perform well for both small and large projects. Some perform well only for small projects. Still, some heuristics that perform well for certain types of small projects may not perform well for other projects of comparable size. The implicit network structure based on precedence relationships and path interconnections influences network complexity and, hence, the performance of scheduling heuristics. The complexity of a project network may indicate the degree of effort that has been devoted to planning the project. The better the planning for a project, the lower the complexity of the project network can be expected to be. This is because many of the redundant interrelationships among activities can be identified and eliminated through better planning.

There have been some attempts to quantify the complexity of project networks. Since the structures of projects vary from very simple to very complex, it is

desirable to have a measure of how difficult it will be to schedule a project. Some of the common measures of project network complexity are presented below:
For PERT networks,

$$C = \frac{\left(\text{number of activities}\right)^2}{\left(\text{number of events}\right)}$$

where an event is defined as an end point (or node) of an activity.
For precedence networks,

$$C = \frac{\left(\text{preceding work items}\right)^2}{\left(\text{total number of work items}\right)}$$

The above expressions represent simple measures of the degree of interrelationship of the project network. Another measure uses the following expression:

$$C = \frac{2\left(A - N + 1\right)}{\left(N - 1\right)\left(N - 2\right)}$$

where A is the number of activities and N is the number of nodes in the project network. One complexity measure is defined as the total activity density, D, of a project network.

$$D = \sum_{i=1}^{N} \max\left\{0, \left(p_i - s_i\right)\right\}$$

where N is the number of activities, p_i is the number of predecessor activities for activity i, and s_i is the number of successor activities for activity i. Alternate measures of network complexity often found in the literature are presented in the expressions below:

w_j = a measure of total work content for resource type j
O = a measure of obstruction factor, which is a measure of the ratio of excess resource requirements to total work content
O_{est} = adjusted obstruction per period based on earliest start time schedule
O_{lst} = adjusted obstruction per period based on latest start time schedule
U = a resource utilization factor

$$C = \frac{\text{Number of activities}}{\left(\text{Number of nodes}\right)}$$

$$D = \frac{\text{Sum of job durations}}{\left(\text{Sum of job durations} + \text{Total free slack}\right)}$$

$$w_j = \sum_{i=1}^{N} d_i r_{ij}$$

$$= \sum_{t=1}^{CP} r_{jt}$$

where

d_i = duration of job i

r_{ij} = per-period requirement of resource type j by job i

t = time period

N = number of jobs

CP = original critical path duration

r_{jt} = total resource requirements of resource type j in time period t

$$O = \sum_{j=1}^{M} O_j$$

$$= \sum_{j=1}^{M} \left(\frac{\sum_{t=1}^{CP} \max\left\{0, r_{jt} - A_j\right\}}{w_j} \right)$$

where

O_j = the obstruction factor for resource type j

CP = original critical path duration

A_j = units of resource type j available per period

M = number of different resource types

w_j = total work content for resource type j

r_{jt} = total resource requirements of resource type j in time period t

$$O_{est} = \sum_{j=1}^{m} \left[\frac{\sum_{t=1}^{CP} \max\left\{0, r_{jt(\text{est})} - A_j\right\}}{(M)(CP)} \right]$$

where $r_{jt(\text{est})}$ is the total resource requirements of resource type j in time period t based on earliest start times.

$$O_{\text{lst}} = \sum_{j=1}^{m} \left[\frac{\sum_{t=1}^{CP} \max\left\{0, r_{jt(\text{lst})} - A_j\right\}}{(M)(CP)} \right]$$

where $r_{jt(\text{lst})}$ is the total resource requirements of resource type j in time period t based on latest start times. The measures O_{est} and O_{lst} incorporate the calculation of excess resource requirements adjusted by the number of periods and the number of different resource types.

$$U = \max_j \{f_j\}$$

$$= \max_j \left\{ \frac{w_j}{(CP)(A_j)} \right\}$$

where f_j is the resource utilization factor for resource type j. This measures the ratio of the total work content to the total work initially available. Another comprehensive measure of the complexity of a project network is based on the resource intensity (λ) of the network:

$$\lambda = \frac{p}{d} \left[\left(1 - \frac{1}{L}\right) \sum_{i=1}^{L} t_i + \sum_{j=1}^{R} \left(\frac{\sum_{i=1}^{L} t_i x_{ij}}{Z_j} \right) \right]$$

where
λ = project network complexity
L = number of activities in the network
t_i = expected duration for activity i
R = number of resource types
x_{ij} = units of resource type j required by activity i
Z_j = maximum units of resource type j available
p = maximum number of immediate predecessors in the network
d = PERT duration of the project with no resource constraint

The terms in the expression for the complexity are explained as follows: The maximum number of immediate predecessors, p, is a multiplicative factor that increases the complexity and potential for bottlenecks in a project network. The $(1 - 1/L)$ term is a fractional measure (between 0.0 and 1.0) that indicates the time intensity or work content of the project. As L increases, the quantity $(1 - 1/L)$ increases, and a larger fraction of the total time requirement (sum of t_i) is charged to the network complexity. Conversely, as L decreases, the network complexity decreases proportionately with the total time requirement. The sum of $(t_i x_{ij})$ indicates the time-based consumption of a given resource type j relative to the maximum availability. The term is summed over all the different resource types. Having PERT duration in the denominator helps to express the complexity as a dimensionless quantity by canceling out the time units in the numerator. In addition, it gives the network complexity per unit of total project duration.

In addition to the approaches presented above, organizations often use their own internal qualitative and quantitative assessment methods to judge the complexity of projects. Budgeting requirements are sometimes incorporated into the complexity measure. There is always a debate as to whether or not the complexity of a project can be accurately quantified. There are several quantitative and qualitative factors with unknown interactions that are present in any project network. As a result, any measure of project complexity should be used as a relative measure of comparison rather than as an absolute indication of the difficulty involved in scheduling a given project. Since the performance of a scheduling approach can deteriorate sometimes with the increase in project size, a further comparison of the rules may be done on the basis of a collection of large projects. A major deficiency in the existing measures of project network complexity is that there is a lack of well-designed experiments to compare and verify the effectiveness of the measures. Also, there is usually no guidelines as to whether a complexity measure should be used as an ordinal or a cardinal measure, as is illustrated in the following example.

RELEVANCE OF COMPLEXITY FACTORS

Using surveys, factors were classified regarding their relevancy (Jaifer et al., 2021). Eighteen among them were judged as most critical. The most prominent research study that synthesizes complexity factors in new product development is probably that of M. Bosch-Rekveldt et al. (see Jaifer et al., 2021). The objective of this study was to develop a new framework for project complexity factors in large engineering projects. The proposed framework was developed based on a literature review and new empirical work consisting of 18 interviews about six projects in the process engineering industry. The resulting TOE (technical, organizational, environmental) framework brought a new element to classical system framework. It highlighted the importance of the environment as a contributor to project complexity. In total, 50 complexity factors were gathered from the literature and from interviews and then grouped according to the three predefined classes of complexity. The TOE framework consists of 15 T-elements, 21 O-elements, and 14 E-elements. Subclasses were also defined as follows: Technical complexity (goal, scope, task, experience, and risk), organizational complexity (size, resources, project team, trust and risk), and environmental complexity (stakeholders, location, market conditions, and risk). The risk sub-class is present in all classes and represents the risk and uncertainty side of complexity.

UNCERTAINTY IN EFFORT AND TIME ESTIMATION

Uncertainty has been also cited as a critical factor in new product development performance, especially for its effects on effort and time. As with complexity, the review of the literature confirmed the lack of standard definitions of uncertainty in project context. The project management body of knowledge (PMBOK) definition is unclear regarding the difference between risk and uncertainty while decision theory defines uncertainty as "condition of the environment of the decision maker such as he or she finds it impossible to assign any probabilities whatever to possible outcomes of an event." Information theory qualifies uncertainty as caused by incomplete information. The strong correlation between uncertainty and complexity in product development projects is obvious in the literature. In fact, uncertainty has been integrated as a dimension in many project-complexity framework. For other frameworks, it was unfortunately neglected, marginalized, limited to market uncertainty, or confused with risk.

From a cause-effect perspective, W. ElMaraghy et al. (Jaifer et al., 2021) stated that a complex system is one when uncertainty exists, otherwise, the system is qualified as complicated. M. BoschRekveldt et al. (Jaifer et al., 2021) recognized the strong correlation between concepts as increased uncertainties would contribute to project complexity. According to T. Williams (Jaifer et al., 2021), uncertainty is characterized by the interdependency of elements and by this means, it is a factor of complexity. R. J. Chapman (Jaifer et al., 2021) considers a complex project as one which exhibits a high degree of uncertainty and unpredictability. Many other authors agree with these assessments.

Regarding the importance of uncertainty in product development performance, many research studies focused on uncertainty effects on effort and time. In some views, uncertainty is the main motivation for engineering change, which in turn should result in cost and schedule overruns. Some qualify uncertainty as closely related to project performance measures: cost, time, scope, and quality. They also reported that uncertainty can be regarded as one of the characteristics of the evolution. That is why managing uncertainty is one of the core elements of a firm's improved performance. Significant research works have been done to confirm the correlation between uncertainty and project performance (time and cost). Increased uncertainties would contribute to the project complexity and hence increase the probability of budget and schedule overruns (Jaifer et al., 2021). C. Stockstrom and C. Herstatt (Jaifer et al., 2021) analyzed uncertainty in product development from a planning perspective. They tested and confirmed some hypotheses about the moderating effects of uncertainty on product development performances. These findings from the literature obviously highlight the importance of uncertainty as a driver of product development effort and time.

EFFORT AND TIME DRIVERS RELATED TO PROFICIENCY

Besides the complexity category of drivers that regroups factors with negative impact on effort and time, drivers related to proficiency were also used in the literature. This concerns those drivers that should enable or ease product development tasks and activities and have a positive impact on effort and time. M. F. Jacome and V. Lapinskil (Jaifer et al., 2021) used productivity for their estimation model of design effort in electronic product development. Other drivers were also used in estimation models like team experience and expertise, design team productivity, product development process, use of appropriate design tools, suppliers experience and involvement, and communication process. Furthermore, many research studies about complexity used some proficiency drivers as complexity factors in the sense that the lack of these factors will generate or increase complexity.

Examples of these drivers are experience with technologies, team cooperation and communication, clarity of goals, and goal alignment. It should be noted that "goal alignment" is one of the key aspects of the integration stage of DEJI Systems Model. The learning and knowledge management process was reported as crucial for product development projects with a dynamic nature such as the case for the aerospace sector. Other researches also highlighted the importance of an effective engineering change-management process, along with effective configuration management and effective risk management process for cost and schedule control.

SYNTHESIS AS A COMPONENT OF INTEGRATION

In the view of DEJI Systems Model, synthesis is synonymous with integration. Thus, studies of time and effort align well with integration, as presented by DEJI Systems Model. As a conclusion of the literature review by Jaifer et al. (2021), effort and time drivers were first classified in complexity, size, risk/uncertainty, and proficiency categories. Nevertheless, according to the literature, elements of project size and uncertainty should belong to the complexity side. Indeed, those elements were first defined as separated from complexity, but the concept of complexity was quickly redefined to include both size and uncertainty. Many research studies confirmed this fact by considering size as a complexity dimension and/ or project size factors as elements of complexity. Thereby, effort and time drivers should be classified either in complexity or in proficiency categories. This classification is supported by the finding of Y. Beauregard (Jaifer et al., 2021) studies, which demonstrated that the engineering task effort is a function of the complexity from one side and the proficiency of resources executing the task from the other side. The classification is also compatible with estimation models that regroup drivers of cost in complexity, size, and productivity. H. D. S. Budiono and M. Lassandy (Jaifer et al., 2021) also claimed that cost estimation has a direct relationship to performance and effectiveness and used complexity index as drivers in their model of cost estimation. This means they believe performance and complexity drive the cost of product development.

ELEMENTS FROM THE CASE STUDY

The broad literature review helped identify and structure many elements of the effort and time drivers' framework. It helped define key concepts involved in effort and time estimation as well. Even with targeting publications about aerospace product development, in many cases elements gathered from the literature are not necessarily specific to this sector and might concern projects of different natures. Because of the exploratory character of this research study and the focus of this book, the practical insight should bring some personalized elements.

A case study was performed (Jaifer et al., 2021) regarding an engineering program in the aerospace industry. The case company is of a very large size (more than 1,000 employees) that operates in the aeronautics sector. It develops and produces products integrating several subsystems and components. Despite the fact that one case study will probably not allow covering all drivers of cost and schedule in aerospace, the large size of the program reviewed in the case study, its nature, and the number of participants enriched the framework and brought out or at least highlighted relevant drivers concerning this industry.

The case study was conducted by a team of five researchers and two internal experts in the context of postmortem analysis. The aim was to analyze key issues and factors that contribute to cost and schedule overruns with the objective of defining drivers of effort and time. Besides this objective, the postmortem analysis aimed also at elaborating a lessons-learned database, identifying key issues with suppliers, and analyzing risks management process. However, these latter are out of the scope of this research. Different methodologies were used starting from data analysis, semi-structured interviews, workshops, and an internal survey. A case study protocol was elaborated by the case study team detailing the problem in its context, the objectives, the timeframe, the data collection and management plan, analysis tools, guidelines, and methodologies including undertaken measures and precautions to protect confidentiality and to ensure objectives achievement The case study protocol was validated by the company board.

CASE STUDY DATA COLLECTION AND ANALYSIS

Data collection and analysis followed a data collection plan that defines required data, data sources, data acquisition tools and templates, and statistical analysis to be done. Three levels of data were concerned in the case study: Program data, interviews and workshops' data, and internal survey data. Program data were used as inputs to guide interviews and workshops' discussions. It supported the understanding of cost and schedule overruns issues by interviewees and workshops participants. These data came from a variety of sources: SAP data management, program schedule, legal contract database, test plan and reports, and other internal databases concerning change management and engineering change management. Interviews and workshops' data concerned the list and hierarchy of potential cost and schedule drivers, as well as all relevant comments and observations reported by participants. Survey data concerned the results of drivers' evaluation and their correlation with costs.

SEMI-STRUCTURED INTERVIEWS

As for interviews, 54 semi-structured interviews were conducted with people from different departments and with different statuses: 11 product development experts, 8 supplier agents, and 35 integrators. The selection of participants was based on their role in project planning, contract management, ongoing activities supervision, change management, tests and validation, or simply regarding their participation in some key events throughout project planning and execution timeline. An invitation with request for consent was sent to potential participants that they could accept or decline. The invitation was accompanied with description of the case study objectives and context, as well as the declaration of confidentiality.

Each participant received a notice at least 2 weeks in advance of the interview with personalized data about cost growth, changes, key events during project's timeline, and appropriate data analysis results, all with a brief reminder of the case study objectives. The course of interviews followed a guideline document describing the followings interview's steps:

1) explanation of case study context and expectations from the interviewee
2) review of main data including the chronology of key events and relevant data analysis results
3) open questions about cost and schedule growth causes or drivers
4) closed questions about the causes of main changes and key events leading to cost and schedule growth
5) root causes of identified causes of cost and schedule growth. This helped to identify the hierarchy of cost and schedule overruns drivers
6) synthesis of identified drivers and interview closure

The guideline was updated after the fifth interview to integrate interviewees and case study team members' suggestions. All team members have previous experience with interview animation in a case study context. At least two team members participated in each interview, of which one is an internal expert and one is a researcher. This is to ensure that all interviewees' sayings and their meanings were correctly understood and noted in the interview transcript. This helped mitigate bias due to personal interpretations. Any disagreement among members participating in interviews was resolved during the weekly team meeting. Transcripts were then validated by interviewees. With the support of an internal team of experts, all interviews' findings were analyzed and synthesized using the current reality tree (CRT) tool and Ishikawa diagram to group drivers of cost and schedule by categories and structure their hierarchy.

WORKSHOPS IN THE CASE STUDY

Eleven workshops or focus groups were organized by work package (WP) to discuss and analyze key findings from interviews in order to finalize the drivers' list. The workshops were animated by an internal expert which is also a case study team member. As for interviews, workshops' participants received a notice 2 weeks in advance accompanied by a synthesis of interviews' results in the form of Ishikawa diagram to help animate discussions and ease the integration of new elements. The number of times each driver has been cited in interviews was used by the workshop participants to judge its relevancy and decide on keeping or dropping it out from the final list of drivers. All decisions about dropping out irrelevant drivers, keeping relevant drivers, and adding new elements that emerge during workshops were made by consensus. Workshops also helped to bring consensus about drivers' classification and categories determination. For most WPs, a second iteration (second workshop) was necessary to reach the consensus and validate the final list of drivers.

INTERNAL SURVEY

Once the final list of drivers was identified, an internal survey allowed for the valuation of drivers using a five-level Likert scale. The objective was to confirm the correlation and show the nature of the relationship between drivers and cost. The valuation was made for each readiness-level category. Drivers related to the readiness-level category were used as criteria for the evaluation. The assessment of the readiness level was based on personal judgments of participants and supported by some empirical data. However, to mitigate bias due to personal perceptions, a detailed supporting document describing criteria and evaluation grid for each readiness level was given to participants to help them decide on which Likert level should be given to each readiness level. This helped standardize the valuation procedure and ensure a normalized understanding of readiness levels' definitions.

To highlight the correlation among identified drivers and costs, graphs of readiness level versus costs were drawn for each WP. To draw a graph for a specific WP, the readiness levels are evaluated for each sub-WPs (subsystem development level) related to the WP.

CASE STUDY LIMITATIONS

Some limitations do appear in this case study. Firstly, identified drivers are related to one case study, so, despite the size of this case study and the number of participants, some potential drivers could be missed. Furthermore, the program involved many suppliers that were not integrated into the case study. Their opinions could enrich discussions during interviews and workshops and allow for more relevant drivers to evolve. Finally, results from the case study were not yet used to estimate the cost or effort of a new project/program and then judge effectively the results of the case study. Thereby, any strong statement about the findings of the case study will be claimed by authors, instead, results helped complete the list of drivers found in the literature from practical insights.

SUMMARY OF EFFORT AND TIME DRIVERS IN THE CASE STUDY

Following the literature review and case study results, and thanks to semantic analysis of terminologies and keywords, elements gathered were redefined if needed, in some cases consolidated and then clustered into complexity drivers and proficiency drivers. On a lower level, elements of complexity were classified into three subcategories: technical, organizational, and environmental, as suggested by the literature. This sub-classification is also supported by the results of a survey involving product development experts in the aerospace industry, in which 83% of interviewees agreed with this classification into these three subcategories. The results of this survey support also the classification of uncertainty into technological, environmental, and organizational subcategories. To simplify the presentation and limit the number of subcategories, project size drivers were integrated into appropriate complexity subcategories instead of creating a new one. Likewise, the three classes of uncertainty: technological, organizational, and environmental were incorporated as elements of the corresponding complexity category. We believe this may not reflect the nature of the relationship between uncertainty and complexity and may also lower the relevance of uncertainty as a key concept in aerospace new product development. Indeed, the aim of this preliminary step of framework development was not to highlight the relevance of each element in the framework, nor the relationship among them; instead, it simply focused on presenting the summary of all gathered elements classified in their appropriate categories. The case study framework contains 14 elements of technological complexity drivers, 12 elements of organizational complexity drivers, 8 elements of environmental complexity drivers, and 13 elements of proficiency drivers for a final list of 47 potential drivers of effort and time. One of the main concerns to be addressed about the elements gathered from the literature and case study is the data saturation condition. In subsequent years of the case study, no new drivers have been identified despite the fact that the number of identified drivers did not decrease. This means that recent studies related to effort and time drivers fail to bring about new elements, which indicates data saturation. Furthermore, the case study, conducted between 2016 and 2017, failed to identify new drivers other than those already existing in the literature, which is another sign of data saturation.

RESULTS FROM THE SURVEY

Given the fact that the predefined framework of effort and time drivers resulted in multiple aspects, the identified list of drivers is quite extensive. Using this preliminary framework as a whole in an estimation model seems to be very difficult if not impossible. In fact, parametric modeling, which is the most used method for effort and time estimation, should use only a few factors or drivers. Furthermore, most of the drivers were gathered from the literature and not necessarily specified to the aerospace sector. That is why a refinement of the framework was necessitated. The aim was to evaluate the relevance and priority drivers regarding the reality of aerospace new product development projects through a survey. Thereby, users of the framework can narrow the choice of drivers for their estimation models if needed. For this purpose, a survey involving aerospace product development professionals was conducted. Further details and graphs from this case study can be seen in Jaifer et al. (2021).

The main objective of results analysis was to classify drivers by categories according to their criticality and, thus, independently from the respondents to the survey. A further research study will be undertaken to analyze drivers' criticality regarding company profiles and develop personalized frameworks consequently. Relying on 29 answers, of which 83% agreed with the classification of complexity drivers into technological, organizational, and environmental. The criticality evaluation results are presented in appropriate graphs and diagrams.

IMPLICATIONS FOR SYSTEMS AND ENGINEERING MANAGEMENT

The results of this research have direct impacts on the accuracy of cost and time estimates. Thus, its implications for engineering management in the aerospace industry concern three of the engineering management domains reported in the engineering management body of knowledge (EMBOK): strategic planning domain, financial resources management domain, and project management domain. New product development is a key means by which companies implement their strategy and achieve revenue objectives. Accurate cost and time estimates contribute undoubtedly to product launch success, which in turn contributes to strategy success. Furthermore, the competitive dynamic of the aerospace sector makes the cost leadership strategy one of the critical success factors. Such strategy could not be achieved without controlling cost through the management of cost estimates. As for financial resources management, cost estimate is one of the most critical inputs to the budgeting and pricing processes. Accuracy of cost estimates is also a prerequisite for many critical decisions concerning contract management, funding, financial sources management, and so on. Regarding project management domain, cost and time estimation accuracy ensures the reliability of the cost and time plans, which are key components in the project management plan. Overall, if the cost and time are estimated within a reasonable degree of accuracy, the ability to schedule, estimate resources, control execution, and conduct trade-offs, among others, will become much easier (Jaifer et al., 2021). Finally, the awareness of the relevance of the identified cost and time drivers, especially those related to complexity and uncertainty may support managers in key decision-making. These drivers are not useful exclusively for estimating matters, their assessment might support critical decisions regarding portfolio management, suppliers' management, resources selection, technologies selection, and risk management. It might also help in personalizing or adjusting engineering and managerial practices from one project to another.

CASE STUDY CONCLUSION

This case study presents a new framework for effort and time drivers in aerospace product development. It goes beyond a simple review of most used drivers; instead, it considers relevant concepts influencing effort and time in aerospace product development like complexity and uncertainty. It is believed that mastering uncertainty, complexity, and proficiency drivers is an overriding condition for an accurate estimation of effort and time in aerospace product development projects. Using inductive reasoning by combining elements from the literature strengthened with elements from the case study and survey involving aerospace product development experts, the resulting framework aims to regroup all potential drivers and classify them by categories. It's the first of its category to offer an overview of potential drivers classified by categories. The determination of drivers' relevancy from aerospace product development allows the customization of the framework for this particular sector. The framework may also be applicable to other sectors sharing similar features of aerospace like automotive, defense, and rail industries. Assessing or measuring the value of some drivers may be a subjective process by nature, in which the perceived value based on previous experiences plays an important role. Because of the differences in skills and experiences, people using the framework and assessing some subjective drivers may come to different conclusions, but long-term use of the framework with a mature planning process will lead to more accuracy in estimations of effort and time. This realization of the differences among users with respect to the prevailing operating environment makes this case study align with the tenets of the DEJI Systems Model.

REFERENCES

Badiru, A. B. (1988), "Towards The Standardization of Performance Measures for Project Scheduling Heuristics," *IEEE Transactions on Engineering Management*, Vol. 35, No. 2, May 1988, pp. 82–89.

Badiru, A. B. (2012), "Application of the DEJI Model for Aerospace Product Integration," *Journal of Aviation and Aerospace Perspectives (JAAP)*, Vol. 2, No. 2, pp. 20–34, Fall 2012.

Jaifer, Rabie, Y. Beauregard, and N. Bhuiiyan (2021), "New Framework for Effort and Time Drivers in Aerospace Product Development Projects," *Engineering Management Journal*, Vol. 33, No. 2, pp. 76–95.

10 Case Study of Nigeria's Industrial Development Centers

INTRODUCTION

Every developing and underdeveloped nation aspires to advance through industrial development programs. The nation of Nigeria is not an exception to that rule. The accounts presented here are based on the author's own personal experience with Nigeria's Industrial Development Centers (IDCs) in the 1990s and public reports about the IDCs over the years. The system failure that occurred with the IDCs is the central theme of the case study covered in this chapter. The chapter is in three major segments, following the framework of the DEJI Systems Model®.

First, a general discussion is presented on why manufacturing is an essential component of the industrialization of a nation. This is the aspired "design" stage of the system. If manufacturing is integrated into the national planning of a nation, in the way advocated by DEJI Systems Model, then industrialization can take hold and it could be sustainable. The absence of integration means that the aspired industrialization would be stunted and would soon fizzle out. This is what happened in the case of Nigeria's IDCs.

Second, the specific case of Nigeria's industrialization efforts via the IDCs is presented. This represents the overlapping stages of "evaluation" and "justification" stages.

Finally, an investigative report on the gradual decline and failure of the IDCs is presented. This highlights what should have been the "integration" stage of the system. The lack of sustainable integration is the major reason that things fell apart for the highly touted IDCs. A summary is presented on how the DEJI Systems Model, if applied rigorously, could help avert a repeat of the fate of the failed IDCs.

SEGMENT 1: INTEGRATED INDUSTRIALIZATION THROUGH MANUFACTURING

Manufacturing is a key foundation for national development. Industrialization is one of the primary means of improving the standard of living in a nation. Manufacturing provides the foundation for a sustainable industrialization. Manufacturing refers to the activities and processes geared toward the production of consumer products. Industrialization refers to the broad collection of activities,

DOI: 10.1201/9781003175797-10

services, resources, and infrastructure needed to support business and industry. History indicates the profound effect that the industrial revolution had on world development. Industrialization will continue to play an active role in national economic strategies. Nations that continue to seek international aid for one thing or another can lay a solid foundation for industrialization through manufacturing enterprises. A nation that cannot institute and sustain manufacturing will be politically delinquent and economically retarded in the long run. A good industrial foundation can positively drive the political and economic processes in any nation (ASCE, 2008; Badiru, 1993; Badiru, 2016; Badiru et al., 2019).

In order to achieve and sustain manufacturing development, both the technical and managerial aspects must come into play. This book contains the managerial processes necessary to facilitate manufacturing development. Project management and technology management are presented as viable means of achieving manufacturing development. Technology transfer is a key component of technology management.

COMPLEXITY OF MANUFACTURING

The economic and social problems facing many nations in the present technological age are very challenging. The gloomy trends have been characterized by stagnant state of established industries, decline in productivity, closure of poorly managed corporations, globalization of markets, increased dependency on physical technology, increased apathy toward social issues, proliferation of organized and sophisticated illegal financial deals, requirement for expensive capital investments, and neglect of economic diversification endeavors. All these problems led to the decline of manufacturing in the 2000s. Fortunately, a resurgence of manufacturing is presently being experienced in different parts of the world.

An industrial development project is a complex undertaking that crosses several fields of endeavors. The diverse political, social, cultural, technical, organizational, and economic issues that intermingle in industrial development compound the manufacturing efforts. Sophisticated managerial approaches are needed to control the interaction effects of these issues. Financial power is a necessary but not sufficient requirement for industrial development. A common mistake by a developing economy is to simply dump scarce financial resources at a development problem without making adequate improvement in the management processes needed to support the development goal. Focusing on the technical aspects of a development project to the detriment of the managerial aspects will only create the potential for failure. A project management approach can facilitate an integrated understanding of the complex issues involved in industrial development and, thus, pave the way for success.

Contrary to general belief, industrial development problems are not limited to the developing nations alone. Even in developed and industrialized nations, large pockets of industrially neglected communities can be found. Residents of these communities live in abject poverty despite the ambient affluence of the overall nation. The countries that are most blessed with natural resources are often those that suffer most from industrial neglect. The problem is typically that they do not know how to initiate and implement the projects that are needed to exploit the available resources. Some rely on misaligned technology transfers. Technology that is aligned with national needs is essential for achieving sustainable development goals. If technology fidelity is high and alignment is high, then we have the greatest potential for success. Proper applications of project management techniques can facilitate further responsiveness and effectiveness to pave the way for the advancement of manufacturing.

INDUSTRIAL DEVELOPMENT AND ECONOMIC DEVELOPMENT

Industrialization is one side of the economic coin. Industrial development can directly translate to economic development if proper management practices are followed. Industrial development can be formulated as a basic foundation for economic vitality and national productivity improvement. Some of the major factors that can positively impact interactions of industrial and economic processes include the following.

- Unification of national priorities
- Diversification of the economic strategy
- Strong strategy for development of rural areas
- Adequate investment in research and development
- Stable infrastructure to support development effort
- Political stability that prevents economic disruption
- Social and cultural standards that improve productivity

Industrial development plays different types of roles in economic development. The structure of certain nations is such that the economic and industrial systems are on different development tracks. To facilitate the interaction of industrial and economic endeavors, plans must be made to allow the industrial system to coexist symbiotically with other production forces.

PURSUIT OF TECHNOLOGICAL CHANGE

Technological change is needed to drive industrial development. Technological progress calls for the use of technology. Technology, in the form of information, equipment, and knowledge, can be used productively and effectively to lower production costs, improve service, improve product quality, and generate higher output levels. The information and knowledge involved in technological progress include those which improve the performance of management, labor, and raw materials.

Technological progress plays a vital role in improving total productivity. Statistics on developed countries, such as the United States, show that in the period 1870–1957, 90% of the rise in real output per man-hour can be attributed to technical progress. It has been shown that industrial or economic growth is dependent on improvements in technical capabilities as well as on increases in the amount of the conventional factors of capital and labor. Technological change is not necessarily defined by a move toward the most modern capital equipment. Rather, technological changes should be designed to occur through improvements in the efficiency of the use of existing equipment. The challenge to developing countries, such as those in Africa, is how to develop the infrastructure that promotes and utilizes available technological resources.

SOCIAL CHANGE AND DEVELOPMENT

Industrial development requires change. A society must be prepared for change in order to take advantage of new industrial manufacturing opportunities. Efforts that support industrial development must be instituted into every aspect of everything that the society does. If a society is better prepared for change, then positive changes can be achieved. Industrial development requires an increasingly larger domestic market. The social systems that make up such markets must be carefully coordinated. The socio-economic impact of industrialization cannot be overlooked. Social changes are necessary to support industrial development efforts. Social discipline and dedication must be instilled in the society to make industrial changes possible. The roles of the members of a society in terms of being responsible consumers and producers of industrial products must be outlined. People must be convinced of the importance of the contribution of each individual whether that individual is acting as a consumer or as a producer. Industrial consumers have become so choosy that they no longer will simply accept whatever is offered in the marketplace. In cases where social dictum directs consumers to behave in ways not conducive to industrial development, changes must be instituted. If necessary, acquired taste must be developed to like and accept the products of local industry.

To facilitate consumer acceptance, the quality of domestic industrial products must be improved to competitive standards. New technologies that facilitate increased industrial productivity must be embraced whether through domestic sources or through technology transfer. For an industrial product to satisfy the sophisticated taste of the modern consumer, it must exhibit a high level of quality and responsiveness to the needs of the consumer. Only high-quality industrial products and services can survive the increasing competitive market. Some of the approaches for preparing a society for industrial development changes are listed below.

- Highlight the benefits of industrial development.
- Keep citizens informed of the impending changes.
- Get citizen groups involved in the decision process.
- Promote industrial change as a transition to a better society.
- Allay the fears about potential loss of jobs due to industrial automation.
- Emphasize the job opportunities to be created by industrial development.

Stipulation for the use of local raw materials is one policy that may necessitate a change in social attitudes. The society (consumers and producers) must understand the importance of local raw materials in domestic manufacturing activities. When backed against the wall of production, producers must develop the necessary indigenous technology or processes for using local materials. The technology transfer modes presented in this book provide a guideline for local adaptation of inwardly transferred technology.

EDUCATION OF PRODUCERS AND CONSUMERS

A more aware population is a more responsive and supportive population. It is essential to include education, training, and public awareness into the strategic planning for manufacturing development. For example, inflation can adversely affect investments in manufacturing development. Inflation is a disease that feeds on itself, but not without the help of people. The basic attitude of citizens toward consumption and wealth greatly determines the trend of inflation. Everyone wants to buy and sell consumer products without paying attention to the production efforts needed to generate those products. The seller wants to sell at whatever exorbitant prices he or she can get. The naive buyer does not help the situation either. He or she is willing to buy whatever he or she wants at whatever the asking price even if it means sacrificing economic personal economic sense, which, over the long run, with many like-minded consumers, can jeopardize the economic path of the whole nation.

The producer, encouraged by the high demand, shrewdly cuts production and, thus, reduces market supply, which then elevates consumer prices even further. The fact that a product is in short supply makes the consumer yearn for more of it. The product quickly becomes a status symbol. The few who can get it claim to be the "class" of the society, and they are willing to pay whatever price to maintain that status quo. Each consumptive individual is unaware of the economic agony that reckless personal acts can cause for the larger population. It is not that consumers are unpatriotic. Rather, it is just that they don't know the adverse implications of collective personal economic habits. This is why educating the public is an important aspect of combating inflation and paving the way for manufacturing development. The producer, the retailer, and the buyer all need to know the real causes of inflation, its boomerang effects, and its potential remedies. The government should play a very active role in the education process. The indoctrination process needs to start right at the grassroots. Children should be exposed to the concepts of battling inflation so that they can grow up to become responsible producers, retailers, or consumers. The early instillation of these concepts to the youths will even have the beneficial side effect of imbibed national pride, which should serve the social and political interests of the nation in later years. Most of the leadership problems in many countries can be traced to a lack of deep-rooted pride and interest in the national welfare.

In an underdeveloped nation, nothing is more discouraging than to hear people proclaim how they have given up on their nation. They claim that nothing works, that no improvements are forthcoming, and that the government is unresponsive to the suffering of the people. How do we expect citizens who feel negative about their nation's performance when they, themselves, reach the position of national leadership? Certainly, attitudes and perception will not change overnight. If citizens carry negative impressions into higher offices, all they will do is to perpetrate their views by saying "didn't I say nothing works in this country?" Obviously, such people have not had early mechanisms of instilling national pride and economic dedication in them. This is why early education and public awareness of economic processes are essential for creating sustainable manufacturing activities.

Government programs cannot succeed without the people's cooperation in the implementation process. Herein lies the irony of governing and being governed. We blame the government for not instituting corrective programs and we turn right around to take actions that innocently sabotage the prevailing programs, thereby compounding the burden of the government. People must be educated on the implications of their actions. Presenting a national strategy is not enough. Ensuring that citizens are educated about how to support and execute the strategy is equally essential. When a consumer product is in short supply, everyone complains and blames the system. No one reflects on the individual contribution at each consumer or producer level. Consider the case of a factory worker who buys a locally made product and complains of its inferior quality compared to imported brands. He or she forgets the personal unique position to contribute significantly to the quality of local products through on-the-job activities. There are many cases of auto industry workers buying foreign-made vehicles not because of the need for diversity of assets but rather due to the belief of the better quality of imported vehicles. This is like sending a signal that says "I build the local vehicle and I know about it. Please don't buy it!" That is a fallacy of production dedication. The worker bites the hand that provides the feeding.

To further illustrate the problem of lost dedication, consider the hypothetical scenario of an auto industry employee. As the demand for imported vehicles increases so do the prices of the vehicles. So, the employee asks the employer for higher wages so that a higher-end imported vehicle can be purchased. In order to pay those higher prices, the employer raises the prices of the locally made vehicles to make more money to pay the higher wages. When the prices of the local vehicles go up without a comparable increase in quality (thanks to the employee's work delinquency), the public buys even less of the locally made vehicles. So, the employer sells fewer cars and makes less money. In the end, the employee is laid off. This shows that the employee's economic education is lacking in the first place. This is why it is essential to broaden the sense of economic awareness of consumers.

The scenarios painted here demonstrate the fact that the economic system needed to support manufacturing development is very complex. Individual actions, as innocent and minor as they may appear, can go a long way in advancing or impeding the overall economy. We should be convinced that the education of the public is a key factor in solving many economic problems before industrial development can take root.

POLITICAL STABILITY AND MANUFACTURING DEVELOPMENT

Politics determines the crux of economic vitality and economic vitality determines the crux of politics. With the changing political environments in developing nations, as those in Africa, it is difficult to stabilize government policies. Many governments are perpetually in transient states. Just when policies are beginning to take a firm hold, a new administration comes in and, as a show of new power, overturns all previous achievements. Billions of dollars have been wasted by succeeding governments through the absurd practice of abandoning projects started by preceding administrations regardless of merit. For the sake of national progress, governments should make a commitment to retain and execute worthwhile projects irrespective of who started them. Foresight should also be exercised when embarking on new industrial projects to ensure that subsequent governments can find the merit for their continuation. Laying a solid foundation for an industrial development project can facilitate a lasting coexistence of political, economic, and industrial activities. The major role of government in industrial development must be that of a facilitator. This is particularly important in inter-government negotiations for mutual development. There is often an interdependence of the political, economic, and industrial systems in every nation. Manufacturing vitality advances along the manufacturing system axis depending on the advances in the economic and political systems. The building blocks of industrial performance are represented by the industrial performance metrics and are dependent on the positive interplay of manufacturing, economic, and political systems.

If the economic and political processes are at low levels, no significant industrial development can be expected. As the economy picks up and the political atmosphere improves, the level of industrial development will begin to increase. To maximize industrial output, the political and economic systems must operate at symbiotically high levels. This is why a developing nation must recognize that chaotic political situations will adversely affect overall national development efforts.

GLOBAL INFLUENCES ON DOMESTIC MANUFACTURING

The new world economic order calls for new approaches to development. The prevailing changes around the world will have profound effects on international trade and domestic manufacturing activities. The changes in Eastern Europe, the advancements in Western Europe, the merging of nations, the breakup of nations, and the emergence of Africa as a viable market will all affect international trade in the coming years. No nation can be insulated from what goes on in other nations. If not impacted directly, impacts can occur through intermediary trading partners. The mobility of manufacturing outputs and products will be one common basis for global trade communication. Companies and countries must recognize the trend and refocus efforts and direct investments accordingly. Some of the key aspects of globalization that impact industrial development activities include the following.

- Transition of some countries from being trade partners to being trade competitors
- Reduction of production cycle time to keep up with the multilateral introduction of new products around the world
- Increased efforts to reduce product life cycle from years to months
- Increased responsiveness to the needs of a mixed workforce
- The pressure to mitigate the adverse impacts of cultural barriers
- Relaxation and expansion of trade boundaries
- Tightening of trade relationships
- Politically driven trade sanctions
- Integration of operations and services and consolidation of efforts
- More effective and responsive communication, including social media
- Increased pressure for multinational cooperation
- Need for multi-company and multi-product coordination
- National security barriers

The interdependence of world projects in modern times makes it imperative that global development efforts be pursued. Political, industrial, economic, and social disasters in one nation can easily spill over to other nations and create adverse chain reactions. Refugee problems now plaguing many nations are cruel reminders that nations can no longer exist in isolation.

There are several unique factors that impinge on industrialization. Some of the factors are addressed in the sections that follow. While technology continues to push the limit of perfection in developed nations, citizens of underdeveloped nations continue to be relegated to antiquated production tools. Industrial development has unique aspects that must be considered in formulating development strategies. Some of these aspects are discussed below.

RURAL AREA DEVELOPMENT FOR INDUSTRIALIZATION

Because of their underdeveloped state, rural areas offer significant potential for industrial development. Ironically, it is that same underdevelopment that makes them susceptible to neglect. Productive members of rural areas move to urban areas in large numbers in search of a better life. The more the technological advancement in a nation, the more the rural residents migrate to urban areas. Consequently, the manpower needed to support the development of a rural area is often not consistently available. The key to keeping the manpower in their localities is the development of the localities and the creation of employment opportunities and the provision of basic amenities of life. Rural neglect has created highly uneven distribution of population in most countries. Urbanization without industrialization is doomed to economic failure.

The sparse population in rural areas means large land areas with potential for industrial setups are available at relatively low cost. This makes them particularly attractive for the location of new plants. But investors often shy away from the rural areas because of the limited availability of supporting infrastructure such as electrical power, water, communication, raw materials, and transportation facilities. If power, water, transportation, and communication are guaranteed, the development of the rural areas will be effected successfully. In addition to assuring these services, a nation should offer incentives (e.g., tax incentives) to rural investors. Some key elements of rural area development include the following.

- Improvement of basic infrastructure
- Creation of investment incentives for local businesses
- Establishment of academic institutions geared to local educational needs
- Establishment of technology extension services through academic institutions
- Creation of a local clearing house for technology information
- Government assistance for local product development
- Provision of adequate rural health care services
- Cultural barriers to industrialization

Cultural barriers can inhibit industrial development. Most industrialization programs are fueled by foreign ideas, technologies, and approaches. In some cultures, cloaks of protection have been developed to protect the society against foreign influences. The greatest fear often involves the potential moral decadence that may accompany industrial development as people migrate from one nation to another in response to the development process. Cultural barriers may also prohibit certain segments of a society from participating fully in the development process. Such cultural barriers must be removed to pave the way for full participative industrial development. Some cultures have evolved to a level of indifference to fraudulent and corrupt practices that are detrimental to development efforts. In some cases, corruption is viewed as an accepted way of doing business. Culturally permissive attitudes to corruption must be altered before general development can take hold. Loyalty to the overall development goals rather than individual pursuits is required as a part of cultural changes for industrial development.

EDUCATION AND INDUSTRIAL DEVELOPMENT

He who owns the knowledge controls the power. Both formal and informal education should play a vital role in national industrial development. In the modern day of high technology, adequate education is needed to succeed in any work environment. Even in some industrialized nations, a large percentage of the adult population in neglected communities is functionally illiterate. It used to be that the children of these poverty-stricken communities were needed to drive the wheels of manual labor in local factories. But the present and future industries, with increasing push for automation, will not need much of the services of the labor-intensive workforce. Education geared toward the new industrial direction will be needed to participate actively in industrial development. Poor parents always proclaim that they don't want their children to end up where they did. Yet, no drastic educational efforts are made to ensure that they don't.

RESEARCH AND DEVELOPMENT PARTNERSHIP

Research and education go hand in hand in creating pathways for industrial development. Partnerships between government, university, and industry offer good avenues for addressing the pressing issues of national development. DEJI Systems Model, if embraced, can create a multi-faceted approach to government-university-industry partnerships. There is enough room for each entity's goals. On the side of business and industry, the end goal is to maximize profit and ensure survival. On the side of the university, the primary goal is the quest for new knowledge, technology advancement, and technology transfer. On the part of the government, the goal is to increase the wealth and security of the nation. Specific research-based partnerships can address the multilateral goals. Quantitative tools such as multi-attribute decision modeling are of interest for this purpose.

Knowledge is a sustainable capital. The establishment of a formal process for the interface of institutions of higher learning and industry can be one of the capitals for industrial development. Universities have unique capabilities that can be aligned with industry capabilities to produce symbiotic working relationships. Private industrial research projects must complement public industrial research programs. Academic institutions have a unique capability to generate, learn, and transfer technology to industry. The quest for knowledge in academia can fuel the search for innovative solutions to specific industrial problems. A collaborative industry is a fertile ground for developing prototypes of new academic ideas. Industrial settings are good avenues for practical implementation of technology. Industry-based implementation of university-developed technology can serve as the impetus for further efforts to develop new technology. Technologies that are developed within the academic community mainly for research purposes often languish in the laboratory because of the lack of formal and coordinated mechanisms for practical industrial implementation. The potentials of these technologies go untapped for several reasons including the following.

- The researcher does not know which industry may need the technology.
- Industry is not aware of the technology available in academic institutions.
- There is no coordinated mechanism for technical interface between industry and university groups.

Universities interested in technology development are often hampered by the lack of adequate resources for research and training activities. Industry can help in this regard by providing direct support for industrial groups to address specific industrial problems. The universities also need real problems to work on as projects or case studies. Industry can provide these under a cooperative arrangement. The respective needs and capabilities of universities and industrial establishments can be integrated symbiotically to provide benefits for each group. University courses offered at convenient times for industry employees can create opportunities for university-industry interaction. Class projects for industry employees can be designed to address real-life industrial problems. This will help industry employees to have focused and rewarding projects. Class projects developed in the academic environment can be successfully implemented in actual work environments to provide tangible benefits. With a mutually cooperative interaction, new developments in industry can be brought to the attention of academia while new academic research developments can be tested in industrial settings. The growing proliferation of distance learning (DL) and distance education (DE) programs can be leveraged to spread educational opportunities to more industry.

The role of government should be as the facilitator to provide linkages and programs that enable multi-organizational collaboration. One desirable approach is for the government to establish and fund technology clearing houses. Academic institutions can serve as convenient locations for technology clearing houses. Such clearing houses can be organized to provide up-to-date information for industrial development activities. Specific industrial problems can be studied at the clearing house. The clearing house can serve as a repository for information on various technology tools. The industry would participate in the clearing house through the donation of equipment, funds, and personnel time. The services provided by a clearing house could include a combination of the following.

- Provide consulting services on technology to industry.
- Conduct on-site short courses with practical projects for industry.
- Serve as a technology library for general information.
- Facilitate technology transfer by helping industry identify which technology is appropriate for which problems.
- Provide technology management guidelines that will enable industry to successfully implement new technology in existing operations.
- Expand training opportunities for engineering students and working engineers.

The establishment of centers of excellence for pursuing industrial development-related research is another approach to creating a favorable atmosphere for industry-university interaction. As an example, from 2000 to 2006, the author founded and directed the Industrial Engineering Center for Industrial Development Research (IE-CIDER) within the Department of Industrial Engineering at the University of Tennessee in Knoxville, Tennessee, United States. The center was dedicated to research studies involving the multi-dimensionality of factors and processes affecting the movement of a product from concept stage to the commercialization stage with specific ties to regional economic development, from a DEJI Systems Model perspective. A part of the services of the center was industry location feasibility studies. A similar university research initiative was developed at the University of Oklahoma in the 1990s. It was within that initiative that the OKIE-ROOKIE expert systems software was developed in the School of Industrial Engineering (Badiru, 1991). The catchy name of the software represents the nickname of the State of Oklahoma, "Okie," while the "rookie" part represents the "newcomer" status of new industry attracted to the state. "Okie," generally and admiringly refers to a resident, native, or cultural descendant of Oklahoma, equating to an Oklahoman (similar to New Yorker). "Okie from Muscogee" is a popular phrase in Oklahoma. With all of this said, the treatise here is that new industry is integrated, per DEJI Systems Model, into the local manufacturing needs of the state.

AGRICULTURE AND INDUSTRIALIZATION

A hungry society cannot be an industrially productive society. It is generally believed that an underdeveloped economy is characterized by an agricultural base. Based on this erroneous belief, several developing nations have abandoned their previously solid agricultural base in favor of alternate means of industrialization. The fact is that a strong agricultural base is needed to complement other industrialization efforts. Agriculture, itself, is a viable source of industrialization when we consider the broad definition of industrialization presented earlier in this book. Mechanized farming is a sort of industrialization. If industrialization does not yield immediate benefits, the society will be exposed to the double jeopardy of hunger and material deprivation. Once abandoned, agriculture is a difficult process to regain. Since agricultural processes take several decades to perfect, revitalization of abandoned agriculture may require several decades. Agriculture should play a major role in the foundation for industrial development. The agricultural sector can serve as a viable market for a concomitant industry through supply chain interfaces. It is interesting to note that the agricultural revolution of the past paves the way for the subsequent industrial revolution.

Human history indicates that humans started out as nomad hunters and gatherers, drifting to wherever food could be found. About 12,000 years ago, humans learned to domesticate both plants and animals. This agricultural breakthrough allowed humans to become settlers, thereby spending less time wandering in search of food. More time was, thus, available for pursuing stable and innovative activities, which led to discoveries of better ways of planting and raising animals for food. That initial agricultural discovery eventually paved the way for the agricultural revolution. During the agricultural revolution, mechanical devices, techniques, and storage mechanisms were developed to aid the process of agriculture. These inventions made it possible for more food to be produced by fewer people. The abundance of food meant that more members of the community could spend that time for other pursuits rather than the customary labor-intensive agriculture. Naturally, these other pursuits involved the development and improvement of the tools of agriculture. The extra free time brought on by more efficient agriculture was, thus, used to bring about more technological improvements in agricultural implements. These more advanced agricultural tools led to even more efficient agriculture. The transformation from the digging stick to the metal hoe is a good example of the raw technological innovation of that time. With each technological advance, less time was required for agriculture, thereby, permitting more time for further technological advancements. The advancements in agriculture slowly led to more stable settlement patterns. These patterns led to the emergence of towns and cities. With central settlements away from farmlands, there developed a need for transforming agricultural technology into domicile technology that would support the new city life. The transformed technology was later turned to other productive uses which eventually led to the emergence of the industrial revolution. To this day, the entwined relationships between agriculture and industry can still be seen. Ideally, there should be a strategic integration of agricultural technology, industrial technology, and business technology.

DEVELOPING WORKFORCE FOR INDUSTRIALIZATION

Technical, administrative, and service workforce will be needed to support industrialization initiatives. People make development possible. No matter how technically capable a machine may be, people will still be required to operate or maintain it. Soon after World War II, it was generally believed that physical capital formation was a sufficient basis for development. That view was probably justified at that time because of the role that machinery played during the war. It was not obvious then that machines without a trained and skillful workforce did not constitute a solid basis for development. It has now been realized that human capital is as crucial to development as physical capital. The investment in workforce development must be given a high priority in the development plan. Some of the important aspects of manpower supply analysis for industrial development include the following.

- Level of skills required
- Mobility of the manpower
- The nature and type of skills required
- Retention strategies to reduce brain drain
- Potential for coexistence of people and technology
- Continuing education to facilitate adaptability to technology changes
- Who is available to educate and train the workforce

With respect to who may be available to contribute to the education and training of the national workforce, the author offers the following perspectives. Military service has an impact in advancing workforce education. By virtue of having gone through structured education and training for their service requirements, ex-servicemen and ex-servicewomen do have unique skills and experience that can be channeled toward educating and training the national workforce for specific needs. Military education can have far more impact on national advancement than we realize. In the days of the author's own university studies in the United States, he marveled at the knowledge base, span of expertise, and professionalism of many of the engineering instructors. He wondered how such a high concentration of marvelous militarily experienced engineers could be found in one institution. He later found out that this was not an isolated incident in one institution. It turned out that a large number of engineering professors in the 1960s and 1970s across the United States had served in the U.S. Navy or Army during World War II. After the war, through government programs, many transferred their military training, education, and expertise into lecturing at universities. The positive impression that the author had of his engineering professors gave him the early incentives to apply more forthright efforts to his engineering education and, subsequently, chose academia as his own career path. The consequence is that the foundational knowledge acquired from the military-engineers-turned professors continues to serve the author's own students years later. The conclusion is that the military directly and indirectly influenced the advancement of technical workforce in the United States. What the United States is experiencing today in terms of being a world leader is predicated on a foundation of consistent technical education over the years. Other nations, particularly developing nations, can learn from this example.

There is often a debate whether the military should continue to invest (and how much) in advanced military education. The fact is that the national investment in advanced military education is not only essential to keep the military on the cutting edge of warfare technology but also to positively impact the national landscape of education that may contribute to the advancement of manufacturing technology. Recognizing the urgent need to address global societal issues from a technical standpoint, in 2008, the National Academy of Engineering (NAE, 2008) of the United States published the "14 Grand Challenges for Engineering." The challenges have global implications for everyone, not just the engineering professions. As such, solution strategies must embrace all disciplines. The military, by virtue of its global presence and wider span of involvement in advanced technical education, can provide the technical foundation for addressing many of the challenges. Engineers of the future will need diverse skills to tackle the multitude of issues and factors involved in adequately and successfully addressing the challenges. Military engineers, in particular, are needed to provide the diverse array of technical expertise, discipline, and professionalism required. Technical education, covering Science, Technology, Engineering, and Mathematics (STEM) at advanced levels, provides a sustainable opportunity for the military to impact the 14 grand challenges listed below.

1. Make solar energy economical
2. Provide energy from fusion
3. Develop carbon sequestration methods
4. Manage the nitrogen cycle
5. Provide access to clean water

6. Restore and improve urban infrastructure
7. Advance health informatics
8. Engineer better medicines
9. Reverse-engineer the brain
10. Prevent nuclear terror
11. Secure cyberspace
12. Enhance virtual reality
13. Advance personalized learning
14. Engineer the tools of scientific discovery

All of the above have implementation implications for manufacturing technology, either directly or indirectly through transitional interactions. An extract from the NAE document on the 14 grand challenges reads:

> In sum, governmental and institutional, political and economic, and personal and social barriers will repeatedly arise to impede the pursuit of solutions to problems. As they have throughout history, engineers will have to integrate their methods and solutions with the goals and desires of all society's members.

Who is better capable to analyze, synthesize, and integrate than the technically trained military? Advanced military education is needed and should be sustained to carry out this charge. A recommendation is that qualified ex-service people should be sent back to school for advanced technical education and turn them into future engineering educators so that future engineering students can benefit and the nation as a whole can also benefit. This may even have the side benefit of having the society view the military in the positive light of national economic advancement beyond national security.

TECHNOLOGY ATTRIBUTES FOR INDUSTRIALIZATION

Some specific attributes of technology can facilitate industrial development. But technology must be managed properly to play an effective role. There is a multitude of new technologies that have emerged in recent years. Hard and soft technologies such as computing tools, cellular manufacturing, intelligent software tools, and social media are changing the landscape of manufacturing rapidly. But much more remains to be done in actual implementation. It is important to consider the peculiar characteristics of a new technology before establishing adoption and implementation strategies for applications in industrial development. The justification for the adoption of a new technology should be a combination of several factors rather than a single characteristic of the technology. The important characteristics to consider include productivity improvement, improved product quality, reduction in production cost, flexibility, reliability, and safety.

An integrated evaluation must be performed to ensure that a proposed technology is justified economically and technically. The scope and goals of the proposed technology must be established right from the beginning of an industrialization project. This entails the comparison of industry objectives with the overall national goals in the areas discussed below.

Market target: This should identify the customers of the proposed technology. It should also address items such as market cost of the proposed product, assessment of the competition, and the market share.

Growth potential: This should address short-range expectations, long-range expectations, future competitiveness, future capability, and prevailing size and strength of the competition.

Impact on national goals: Any prospective technology must be evaluated in terms of the direct and indirect benefits to be generated by the technology. These may include product price versus value, increase in international trade, improved standard of living, cleaner environment, safer workplace, and improved productivity.

Profitability: An analysis of how the technology will contribute to profitability should consider past performance of the technology, incremental benefits of the new technology versus conventional technology, and value added by the new technology.

Capital investment: Comprehensive economic analysis should play a significant role in the technology assessment process. This may cover an evaluation of fixed and sunk costs, cost of obsolescence, maintenance requirements, recurring costs, installation cost, space requirement cost, capital substitution potentials, return on investment, tax implications, cost of capital, and other concurrent projects.

Skill and resource requirements: The utilization of resources (manpower and equipment) in the pre-technology and post-technology phases of industrialization should be assessed. This may be based on material input-output flows, the high value of equipment versus productivity improvement, the required inputs for the technology, the expected output of the technology, and the utilization of technical and nontechnical personnel.

Risk exposure: Uncertainty is a reality in technology adoption efforts. Uncertainty will need to be assessed for the initial investment, return on investment, payback period, public reactions, environmental impact, and volatility of the technology.

National productivity improvement: An analysis of how the technology may contribute to national productivity may be verified by studying industrial throughput, efficiency of production processes, utilization of raw materials, equipment maintenance, absenteeism, learning rate, and design-to-production cycle.

FOUR-SIDED INFRASTRUCTURE FOR INDUSTRIALIZATION

Industrialization built upon a solid foundation can hardly fail. The major ingredients for durable industrial, economic, and technological developments are electrical power, water, transportation, and communication facilities. These items should have priority in major industrial development projects. Of course, housing infrastructure for the workforce is another basic need that must be accounted for in an industrial development strategy. A four-sided manufacturing infrastructure template to provide a foundation for industrial development will have four primary elements, as summarized below:

1. Reliable power supply
2. Consistent water supply
3. Good transportation system
4. Efficient communication system

The supporting amenities will consist of the following desirables:

- Housing
- Education
- Health care

The provision of adequate health care facilities is particularly essential to building a strong industrial base. A healthy society is a productive society while an unhealthy society will be an economically destitute society. Diseases that often ravage impoverished nations can curtail the productive capabilities of the citizens. Recent global problems with contagious diseases, such as the Bird Flu, SARS, Avian Flu, Enterovirus, and Ebola, have deleterious effects on the productive potentials of the nations affected. The destructive effects of pandemic and epidemic infectious diseases can be mitigated by prompt access to basic health care services. With fast global human interfaces nowadays, no nation can be immune to infectious diseases across the continents.

RELIABLE POWER SUPPLY

Electricity is the major source of power for production facilities. This fact has been realized by developing countries for a long time. Yet, not enough effort has been directed at adequate generation and reliable distribution of this very important resource. Where electricity has been generated in abundant quantities, reliable distribution has been miserably lacking. Plants have been shortsightedly constructed without planning for adequate power supply. No wonder then that most industries in developing economies are running only at a fraction of their capacities. The problems plaguing power supply companies should be critically studied so that a lasting solution can be found. A large portion of the initial development efforts should be directed at ensuring adequate power supply. Once supply is found to be reliable and stable, major production endeavors can then be pursued.

CONSISTENT WATER SUPPLY

In today's technology, chemicals play a significant role in product development. Water is an indispensable component of the use of chemicals in industrial operations. Water is needed to not only sustain life but to also support the many products that make life livable. Water facility development should be addressed in the two categories below.

- Industrial water
- Potable water

Because of its huge volume of demand, industry should not compete with households for water supply. Like electricity, reliable water supply should be assured before urging investment in production facilities. There are cases of where a new industry opens, has water supply for a short period, and then is disappointed by the reality of water shortages.

ACCESSIBLE TRANSPORTATION SYSTEM

In agreement with Newton's Law of Motion, nothing moves without transportation. In the ancient days of geographically limited commerce, rapid transportation, though important, was less of a concern. But in the modern society, the global market has necessitated interactions with far-away locations. Developed economies appreciate the necessity of these interactions, and they developed transportation systems to meet the needs. Developing economies are "still developing" because their transportation systems, in many respects, are yet to catch up with the rest of the world. Progress in modern commerce depends on the facilities for conveying products from one location to another efficiently, reliably, and safely.

After power and water, transportation should have the third priority in any strategic development plan for manufacturing development. Good roads should be constructed and maintained to provide a lasting support for development efforts. Industrial access roads as well as workforce access roads should be constructed. The economic loss can be enormous if the transportation system is neglected even in a developed economy. For example, The American Society of Civil Engineers (ASCE) reported that 42% of America's major urban highways remain congested, costing the economy an estimated $101 billion in wasted time and fuel annually (ASCE, 2013). The report confirms further that the U.S. Federal Highway Administration estimates that $170 billion in capital investment would be needed on an annual basis to significantly improve conditions and performance of the infrastructure. This is a huge financial requirement even for a wealthy nation. For developing nations, this book suggests incremental development, sustainment, and investment in transportation infrastructure. Little blocks of building are more manageable and sustainable. It should be understood that the development of a good transportation system can take a very long time at a very high cost. With its accessibility, the U.S. interstate system has fueled development in several parts of the United States.

EFFICIENT COMMUNICATION SYSTEM

Communication propels commerce. Telephone, the Internet, and the postal system constitute alternate means of conducting business. In a developing economy, the lack of a reliable telephone service or reliable online systems can force entrepreneurs to make physical appearances when conducting even the most basic of business transactions. This, of course, necessitates the increased use of roads, thereby, overloading the already inadequate transportation system. In an underdeveloped community, business that can be conveniently conducted over the telephone is normally conducted by physical presence. The requirement for physical mobility directly takes a person from productive activities, consequently creating adverse effects on national productivity.

In addition, frequent road transportation increases the exposure of the traveler to the hazards of the precarious transportation system. The more the people need to be on the roads, the higher the congestion, and the greater the risks of road mishaps, which have been known to rob a nation of some of its most productive citizens. The reliability of electric power, water, transportation, and communication services is essential for a sound industrial development. Nations with well-developed communication systems have used them to further their development programs. On the other hand, those who have not paid proper attention to their communication systems will continue to struggle with dormant or regressing economies.

INDIGENOUS ECONOMIC MODELS

The formulation of an indigenous economic model is relevant for industrial development. The unique aspects of a developing economy that will necessitate indigenous models are discussed in this section. The reasons why imported economic principles and models may not work in a developing nation are presented. The success of indigenous economic models in an underdeveloped area can serve as a useful paradigm for industrial development endeavors in nations with similar economic plights. A synergistic implementation of indigenous economic models will have a positive impact on the overall world economy. Both the economic and non-economic aspects of industrial development must be accounted for in developing an indigenous model. Well-founded laws and principles of economics have been utilized in the process of formulating existing economic policies. Unfortunately, none of the policies has yielded totally satisfactory results in underdeveloped nations. There has been no short-term success, and there is no guarantee of a long-term success. Restrictions on importation, barter agreements, structural adjustment of currency value, and other economic policies are just a few examples of the concerted efforts being made by some developing countries. The fact that some of these efforts have been or may be unsuccessful indicates that there are certain aspects of the domestic economic situation that are being overlooked, oversimplified, or unrecognized in a developing nation.

The standard economic models that are widely employed throughout the world were developed based on societal behaviors. Many emerging economies, with their peculiarities, simply don't fit the formulation of those models. That, perhaps, is why the economies in those societies have not responded positively to the traditional economic policies. Let us consider the laws of supply and demand for example. The laws are based on the assumption that consumers and producers will respond to certain inputs in some rational fashion. The downward-sloping demand curve indicates that consumers will buy less of a product as the price is increased. By analogy, the upward-sloping supply curve indicates that suppliers will supply more of a product as the price increases. This is quite logical, except in certain underdeveloped economies. In many developing societies, the basic supply and demand laws may not be directly applicable. This is because people are expected to behave in manners foreign to their social, cultural, political, and economic structures and attitudes. Rather than try to mold a developing nation to fit existing economic models, attempts should be made to modify the models to fit the unique situations of the nations. Only then can we have models that accurately explain and predict the actions of "developing" consumers and, thus, provide reliable guidelines for effective national economic policies. Atta (1981) constructs a macro-economic model of the Ghanaian economy. The model emphasizes the supply side of a small open economy. This presents a good example of an economic view tailored to an indigenous scenario.

The indigenous banking system that has been established in India is another excellent example of an economic strategy tailored to a unique national need. The indigenous banking system is based on the strength of personal relations, which is a strong social link in India. The personal relations are used to vouch for credit worthiness instead of using the conventional collateral approach. This indigenous banking system has worked very well and it has made it possible for rural people, who would have otherwise not gone to the Western-style banks, to actively engage in indigenous banking transactions. In a developing nation, the familiar curves of supply and demand may be mangled by several factors. Some of these factors are quite obvious. Some, on the other hand, are very subtle and can only be fully identified, understood, and quantified through dedicated economic research. Economic research must be directed specifically at formulating indigenous economic models. Some important factors that should be considered in such a research effort include the following.

- The level of propensity to consume in a developing nation (consumption orientation rather than production orientation
- The inferiority complex that prompts some societies to prefer imported goods and services
- The lack of self-pride in the products of local labor
- The affinity for black-market transactions

The factors above, in addition to others to be determined through appropriate locally focused research, should form the nucleus of an economic model for a developing country. As far back as 1975, the United Nations (UN) initiated a program to ensure that adequate quantitative information is available for the planning of economic and social development in Africa. Today, an indigenous economic model could be a significant component of similar quantitative modeling. It should be recognized that many developing nations are capable of generating their own economic precedents. They typically have the number (in population) and the capability to support indigenous models. Through the practice of hoarding, "developing" retailers already know how to create artificial scarcity. They already know how to mount a social assault on the equilibrium point of the standard supply and demand curves. So, there is nothing like "market price" in existence in many developing nations. Some suppliers have already customized the principles of fair competition to accommodate collusion. Through the forces of collusion, the suppliers can collectively dictate their harsh terms to hapless consumers. In order to lay a serious foundation for industrial development, indigenous economists should be charged with the responsibility of conducting the appropriate research to study the feasibility of indigenous economic models for a nation that is aspiring to be industrialized. This may be effectively accomplished through post-graduate thesis research at the nation's institutions of higher learning. The research may be accomplished under the university-industry cooperative model suggested earlier.

In an in-depth analysis of national development planning, Gharajedaghi (1986) stressed the fact that a plan of action must not be shortsighted by descriptive adjectives that tell us nothing about the nature of the problem being solved. In government functions, there is a tendency to place responsibility for dealing with a problem in that part of government where a relevant functional adjective can be found. For example, if a problem is found in transportation, the blames are directed at the transportation department. The responsibility for dealing with the problem is automatically assumed to be that of the department even though the root of the problem may be a lack of public discipline. Adjectives and nouns used to describe problems (e.g., health, finance, social, economic, and political) often tell us nothing about the real problem being faced. The adjectives simply indicate the point of view of the person looking at the problem. This pitfall must be avoided if a workable indigenous economic model is to be found. All aspects of a nation's economic system (e.g., social, education, and information) must be assessed in the development of the model.

The benefits of having an indigenous economic model are numerous. If successful, it can encourage the country to undertake other self-help projects that might, otherwise, be left to external forces. It can serve as an incentive for citizens to believe in their own economic system. It can force people to exhibit more responsibility in their actions. The nation can exude pride for taking her own destiny in her own hands. The implementation of the model can serve as an example and guideline for other countries in similar economic situations. The economic model can be presented to the citizens as an accurate description of their behaviors. That may compel them to reflect and become more conscious of the way they approach consumption of goods and services, thereby providing avenues for controlling inflation. The model can be an educational tool by making each sector of the economy more appreciative of the rationale behind the actions of other sectors. The idea of an indigenous economic model should provide the directions in which to look for solutions to the pervasive development problems facing the nation. The development of a structure indigenous economic model will be subject to the following categories of factors.

Technology

- Raw materials
- Technology transfer
- Local research
- Skilled labor
- Technology infusion
- Local-context research

Industry Diversity

- Oil and gas
- Farming
- Goods and services
- Import-export transactions
- Industrial production facilities

External Influences

- Foreign markets
- Technology access
- Trade obligations and restrictions
- Laws and regulations

Public infrastructure

- General public response
- Social climate
- Government agencies
- Political system
- Banking system
- National productivity
- Health care system

Education

- Educational system
- Funding sources
- Research engagements
- Study abroad opportunities
- Industry involvement

Products

- Preferences and options
- Quality
- Availability
- Supply and demand

- Cost and inflation
- Brand options

Entrepreneurial opportunities

- Business climate
- Business loan
- Financial resources
- Taxes and incentives
- Business etiquette
- Collusion
- Monopoly

Government

- Bribery and corruption
- Investment climate
- Exchange control
- Immigration system
- External affairs

PRODUCT STANDARDIZATION AND SYSTEMS INTEGRATION

Product standardization is another factor that can facilitate industrial development. If a standard is available, product interchangeability will be possible. Thus, the market for an industrial product can expand. With an expanded market, better levels of industrialization can be achieved. Units of measure have been one area where standardization has been pursued in many nations. For example, conversion to the metric system has been pursued by several nations as a means to facilitate product compatibility and increase world trade. As more and more countries are switching to metric, a standard system of measurements will simplify international trade. The process of standardization is complicated. But once the initial difficulties of adopting a standard have been overcome, industrial exchanges will become easier. One of the major problems in industrial development is the lack of product integration. We may be effective in making good individual products. But when it comes to fitting the products together to arrive at an overall assembled system, we may have difficulties if there are no prevailing standards. Products should be designed with consideration for how they support one another to achieve an overall workable system in the match toward national industrialization.

STRATEGIC PLANNING FOR MANUFACTURING

Advance planning and dedication are important for industrialization. Industrial planning determines the nature of actions and responsibilities required to achieve industrial development goals. Strategic planning involves the long-range aspects of manufacturing development efforts. Planning forms the basis for all actions.

Strategic planning for manufacturing development can be addressed at three distinct levels as discussed below.

Supra-Level Planning

Planning at this level deals with the big picture of how manufacturing development fits the overall and long-range needs of the community, the region, or the nation. Questions faced at this level may concern the potential contributions of manufacturing activities to the standard of living in the community, the development of resources needed to provide basic amenities in the community, the required interfaces between development projects within and outside the community, the government support for manufacturing development, responsiveness of the local culture, and political stability.

Macro-Level Planning

Planning at this level may address the overall planning within a defined industrial boundary. The scope of the development effort and its operational interfaces should be addressed at the macro-level planning. Questions addressed at this level may include industry identification, product definition, project scope, availability of technical manpower, availability of supporting resources, workforce availability, import/export procedures, development policies, effects on residential neighborhoods, project funding and finances, and project coordination strategies.

Micro-Level Planning

This level of planning deals with detailed operational plans at the task levels of manufacturing activities. Definite and explicit tactics for accomplishing specific development objectives should be developed at the micro level. Factors to be considered at the micro-level planning may include scheduled time, training requirement, tools required, task procedures, reporting requirements, and quality requirements. Manufacturing is capital intensive. If not planned properly, the investment may prove to be an uneconomic venture.

SEGMENT 2: EVALUATION AND JUSTIFICATION

THE CASE OF NIGERIA'S IDCs FOR INDUSTRIALIZATION

As the largest economy in Africa, it is appropriate to use case examples from Nigeria to illustrate the potentials for Africa-US technology and industrialization interactions. While on assignment for the United Nations Development Program (UNDP) in the mid-1990s, the author had the privilege of working with the Nigeria Federal Ministry of Industry in Abuja, Nigeria. His specific assignment was on the UNDP program named TOKTEN (Transfer of Knowledge Through Expatriate Nationals). On that assignment, he witnessed and participated in several industrialization initiatives. His specific role was to develop operational guidelines for the Industrial Development Centers (IDCs) located in every state of the country at that time. A key part of his responsibilities for the IDCs was to explore technology

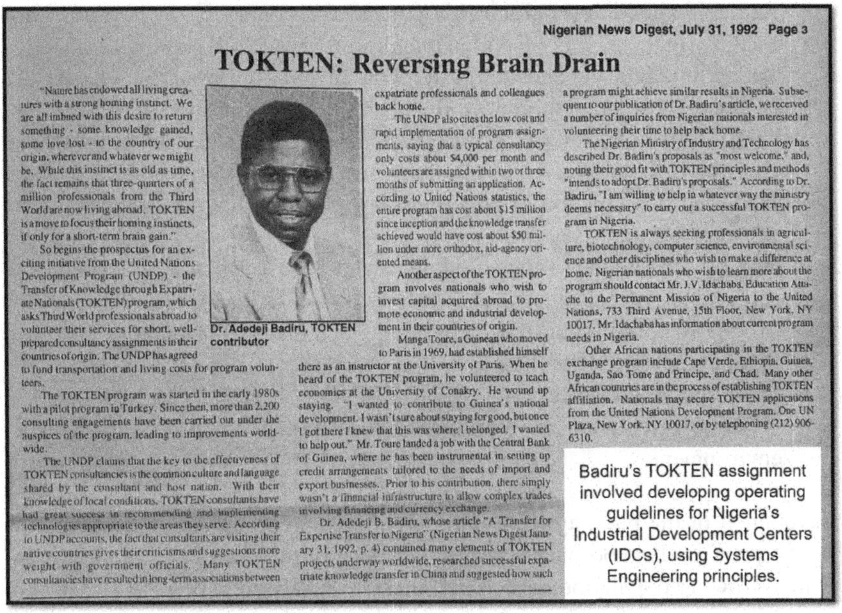

Nigerian News Digest, July 31, 1992 Page 3

TOKTEN: Reversing Brain Drain

"Nature has endowed all living creatures with a strong homing instinct. We are all imbued with this desire to return something - some knowledge gained, some love lost - to the country of our origin, wherever and whatever we might be. While this instinct is as old as time, the fact remains that three-quarters of a million professionals from the Third World are now living abroad. TOKTEN is a move to focus their homing instincts, if only for a short-term brain gain."

So begins the prospectus for an exciting initiative from the United Nations Development Program (UNDP) - the Transfer of Knowledge through Expatriate Nationals (TOKTEN) program, which asks Third World professionals abroad to volunteer their services for short, well-prepared consultancy assignments in their countries of origin. The UNDP has agreed to fund transportation and living costs for program volunteers.

Dr. Adedeji Badiru, TOKTEN contributor

The TOKTEN program was started in the early 1980s with a pilot program in Turkey. Since then, more than 2,200 consulting engagements have been carried out under the auspices of the program, leading to improvements worldwide.

The UNDP claims that the key to the effectiveness of TOKTEN consultancies is the common culture and language shared by the consultant and host nation. With their knowledge of local conditions, TOKTEN consultants have had great success in recommending and implementing technologies appropriate to the areas they serve. According to UNDP accounts, the fact that consultants are visiting their native countries gives their criticisms and suggestions more weight with government officials. Many TOKTEN consultancies have resulted in long-term associations between expatriate professionals and colleagues back home.

The UNDP also cites the low cost and rapid implementation of program assignments, saying that a typical consultancy only costs about $4,000 per month and volunteers are assigned within two or three months of submitting an application. According to United Nations statistics, the entire program has cost about $15 million since inception and the knowledge transfer achieved would have cost about $50 million under more orthodox, aid-agency oriented means.

Another aspect of the TOKTEN program involves nationals who wish to invest capital acquired abroad to promote economic and industrial development in their countries of origin.

Manga Toure, a Guinean who moved to Paris in 1969, had established himself there as an instructor at the University of Paris. When he heard of the TOKTEN program, he volunteered to teach economics at the University of Conakry. He wound up staying. "I wanted to contribute to Guinea's national development. I wasn't sure about staying for good, but once I got there I knew that this was where I belonged. I wanted to help out." Mr. Toure landed a job with the Central Bank of Guinea, where he has been instrumental in setting up credit arrangements tailored to the needs of import and export businesses. Prior to his contribution, there simply wasn't a financial infrastructure to allow complex trades involving financing and currency exchange.

Dr. Adedeji B. Badiru, whose article "A Transfer for Expertise Transfer to Nigeria" (Nigerian News Digest January 31, 1992, p. 4) contained many elements of TOKTEN projects underway worldwide, researched successful expatriate knowledge transfer in China and suggested how such a program might achieve similar results in Nigeria. Subsequent to our publication of Dr. Badiru's article, we received a number of inquiries from Nigerian nationals interested in volunteering their time to help back home.

The Nigerian Ministry of Industry and Technology has described Dr. Badiru's proposals as "most welcome," and, noting their good fit with TOKTEN principles and methods "intends to adopt Dr. Badiru's proposals." According to Dr. Badiru, "I am willing to help in whatever way the ministry deems necessary" to carry out a successful TOKTEN program in Nigeria.

TOKTEN is always seeking professionals in agriculture, biotechnology, computer science, environmental science and other disciplines who wish to make a difference at home. Nigerian nationals who wish to learn more about the program should contact Mr. J.V. Idachaba, Education Attache to the Permanent Mission of Nigeria to the United Nations, 733 Third Avenue, 15th Floor, New York, NY 10017. Mr. Idachaba has information about current program needs in Nigeria.

Other African nations participating in the TOKTEN exchange program include Cape Verde, Ethiopia, Guinea, Uganda, Sao Tome and Principe, and Chad. Many other African countries are in the process of establishing TOKTEN affiliation. Nationals may secure TOKTEN applications from the United Nations Development Program, One UN Plaza, New York, NY 10017, or by telephoning (212) 906-6310.

Badiru's TOKTEN assignment involved developing operating guidelines for Nigeria's Industrial Development Centers (IDCs), using Systems Engineering principles.

FIGURE 10.1 Badiru's 1992 newspaper article on Nigeria's TOKTEN program.

transfer initiatives that could benefit the country's industrial establishments. The accounts presented here are based on Sahel (1992). The accounts confirm that there is no shortage of laudable planning in developing countries. The problems are often that of sustainable implementation of the plans, which is what is addressed by the "integration" stage of the DEJI Systems Model. The fact that the IDCs fell into a shameful state of disrepair confirms that the recommendations presented to UNDP and Nigeria's Ministry of Industry were never carried out, from a systems perspective. This issue points to the need for new management approaches to national planning schemes. It is hoped that the contents of this book can steer policy makers, decision-makers, leaders, practitioners, portfolio managers, and entrepreneurs into the right direction of better management of technology transfer engagements. Figure 10.1 shows the author's 1992 article on the TOKTEN program.

Nigeria's Policy Objectives of the 1980s

The overriding aim of government's development program in the Fourth National Development Plan was improvement in the living conditions of the people. A number of specific objectives were focused on this goal. The specific objectives set for the Fourth Plan period included the following:

- Increase in the real income of the average citizen
- More even distribution of income among individuals and socio-economic groups

- Reduction in the level of unemployment
- Increase in the supply of skilled manpower
- Reduction of the dependence of the economy on a narrow range of activities
- Balanced development, that is, the achievement of balance in the development of the different sectors of the economy and the various geographical areas of the country
- Increased participation of Nigerians in the ownership and management of productive enterprises
- Greater self-reliance—increased dependence on our domestic resources to achieve the various objectives of society. This also implied increased efforts to achieve optimum utilization of human and material resources
- Development of technology
- Increased productivity
- National orientation to discipline, better attitude to work, and cleaner environment

The context of the plan, agricultural production, and processing were given the highest priority. This became necessary because of the need to feed a large and rapidly growing population without the massive importation of food and to provide the basic raw materials needed for agro-based industries. It was also a necessary strategy for the development of the rural areas so as to reduce migration from rural to urban areas. A rapid growth in agricultural productions was an essential component of the strategy of self-reliance which was a major objective of the plan.

Education and manpower development, strengthening of economic infrastructures, and provision of health and housing needs constituted priority areas of the plans.

The Fourth Plan was launched at a time when the country's production of crude oil, which was the main source of government revenue and foreign exchange earnings, had virtually stabilized. As conceived in the previous Third Plan period, the basic strategy was that of using the resources generated by this asset to ensure expansion of the productive capacity of the economy in order to lay a solid foundation for self-sustaining growth and development in the shortest time possible. As part of the strategy for moving the economy toward greater self-reliance, the use of domestic resources, in the planning and execution of projects was encouraged. Government agencies were similarly encouraged where appropriate, to develop in-house capacity for the execution of projects. The growth of indigenous contracting capacity was also given support especially with respect to civil engineering construction and the patronage of Nigerian companies.

NIGERIA'S INDUSTRIALIZATION OBJECTIVES OF THE 1980S

Apart from the emphasis given to the establishment of basic industries and the inclusions of technology development objectives in the third and fourth national development plans, the objectives of industrial development of the 1970s continued into the early 1980s. These objectives included among others providing greater employment opportunities, increased export of manufactured goods, dispersal of industries, improving technological skills and capability, increased local content of industrial output, attracting foreign capital, etc.

The major goal of the government in the industrial sector in the 1980s was to encourage and promote directly and indirectly rapid development of manufacturing and allied activities. In order to ensure that industrialization brought in its wake a truly beneficial economic and social development, industrial development was guided by the following objectives:

Manufacturing value added: Industries with low local inputs into manufacturing contribute minimally to national economic development. In order to encourage meaningful economic development, industrial enterprises which utilize local inputs were encouraged. Particular attention was given to the following:

- Utilization of local raw materials
- Agro-based industries
- Industries with linkage effects
- Industries with backward integration potentials
- Manpower development
- Technology development

Industrial Research and Development (R&D): The continued improvement of the quality of industrial products, the production process, and the amount of local input are desirable attributes which industries should pursue through R&D. In addition to the research institutions established by government, private sector industrial establishments were encouraged to maintain effective R&D units that can find solutions to the problems faced in industries.

Employment: The industrial sector was expected to generate remunerative employment for large numbers of people throughout the country. Within the limits of economic viability, industries were encouraged to adopt labor-intensive technologies. Industries based on materials available in each locality or making the most of local skill and manpower enjoyed more governmental incentives.

Export Promotion: Apart from satisfying domestic demand, Nigerian industries were also expected to produce goods and services for export. To engage meaningfully in international trade meant that Nigerian industrial products should be internationally competitive in terms of both quality and price.

Industrial Dispersal: The dispersal of industries throughout the country was aimed at reducing the concentration of industries in a few areas. It was also designed to check rural-urban migration and promote balanced physical development.With rapid deterioration in the balance of payment since 1982, tariffs were increased, quantitative controls were extended to a wide range of imports, and import licensing and the allocation of foreign exchange were introduced. This highly protective regime insulated domestic industries from foreign competition and encouraged the growth of import-based consumer goods and assembly industries with little domestic value-added.

THE STRUCTURAL ADJUSTMENT PROGRAM AND INDUSTRIALIZATION

The introduction of SAP was the culmination of various attempts to contain the structural defects which became apparent in the economy as a result of the collapse of oil prices in 1981. The fall in oils prices led to a fall in Nigeria's foreign reserve from about N 5.8 billion by mid-1981 to under N 1.2 billion by early 1982. Real GDP fell by 1.9% in 1982 and 6.3% in 1983. Attempts to secure a $2.0 billion standby facility from the IMF and to raise loans through syndication by a consortium of banks failed because "a certificate of good economic health and management" was required. To achieve the requirements of the IMF as far back as 1982, government introduced the 1982 Stabilization Act, which brought a number of austerity measures, including some of the IMF requirements, that is, reduction in budget and balance of payments deficits, reduction in government spending, and cutting subventions to parastatals. However, the government did not agree to a devaluation of the naira.

The Economic Recovery Program announced in the 1986 annual budget had the following major objectives:

- To restructure and diversify the productive base of the economy in order to reduce dependence on the oil sector and imports
- To achieve fiscal and balance of payments viability over the medium term
- To lay the basis for sustainable non-inflationary growth over the medium and long term

In order to achieve the above objectives, the following policy elements were built into the program.

- Strengthening existing demand management policies
- Adoption of a realistic exchange rate policy
- Further rationalization/restructuring of the customs tariffs to aid the promotion of industrial diversification
- Simplification of the regulations and guidelines governing industrial investment and commercial banking activities
- Adoption of appropriate pricing policies especially for petroleum products and public utilities

The 1986 budget, in effect, gave the basic outline of SAP that was introduced later in the year. The core policies introduced involved three main approaches.

- Actions to correct the over-valuation of the naira (introduction of the Second-Tier Foreign Exchange market)
- Actions to overcome the observed public sector inefficiencies (rationalization/privatization and deregulation)
- Actions to relieve the debt-burden and attract a net inflow of foreign capital while keeping a lid on foreign loans

Exchange Rate

Prior to the advent of SAP, the exchange rate of the naira vis-à-vis the major convertible currencies was fixed by the CBN, allowing only for a few minor adjustments. Its allocation among contending demands was by import licensing. Apart from its arbitrariness and administrative rigidity, the import licensing procedure was fraught with corruption and maladministration and led to the mis-allocation of scarce resources.

The introduction of the Second-Tier Foreign Exchange Market (SFEM) brought in a measure of market forces determination of the exchange rate. Between September 1986 and April 1987, there were two foreign exchange markets, the First-Tier, that is, fixed rate to which a "crawling peg was applied" in order to slide it down until the gap between it and the Second-Tier (Second-Tier Foreign Exchange Market (SFEM) was closed.

Three modifications were introduced into the operation of the market:

- The first was the merging of the first and second tiers in April 1987.
- The introduction of the "Dutch Auction" system in April 1987 whereby the banks pay whatever they bid, instead of paying the emerging marginal rate for the day.
- The third shift came in January 1989 which enforced the requirements that autonomous funds be sold at the rate which emerges from the auction.

Trade Policies

Before SAP, import procedures were a maze of regulations. Along with SFEM came the abolition of import licensing. The 30% import duty surcharge introduced under the Emergency Decree was also abolished. Custom duties were streamlined and made more uniform. This culminated in the Customs and Excise Tariff Consolidation Decree of 1988. This Decree gave a 7-year life for specified rates in order to give certainty to business planning. The prohibition list was equally reduced to 16 items.

The devalued naira has enhanced the naira value of traditional non-oil exports. The SFEM Decree allowed non-oil export revenue to be retained 100% by the exporter in his domiciliary account and to convert proceeds in such account into naira through the inter-bank foreign exchange market.

Fiscal Policies

It was expected that in line with the entire adjustment process, government expenditure would be trimmed and budgetary deficits held down to within 3% of GDP. To achieve this, there was to be:

- restraints on growth of public wage bills;
- emphasis on maintenance culture rather than on starting new projects;

- privatization or commercialization of certain parastatals and government owned companies;
- allocation of capital expenditure to key projects that are near completion, and which have greater impact on other sectors; and
- emphasis on market determined prices: subsidies were to be removed, especially those on petroleum and energy.

However, these have not been easy to apply. In 1987, the deficit went up to almost 10% of GDP despite increased naira revenue from oil and apparent increase in domestic taxation. Efforts have, however, been made to implement a number of these measures.

The introduction of SAP in 1986 was inevitable given the palpably dismal situation in the first half of the 1980s. However, the efficacy of the program, given the excruciating demand which it has made on the individual and on social structures, needs to be assessed. In the implementation of the reforms, various targets are set, though they were reviewed periodically.

In 1987 and 1988, the credit ceilings for both the private and public sectors were exceeded by very wide margins. In the succeeding year, however, actual performance was far below the target for both the private and public sectors. Though actual performance in 1990 was higher than the target for both sectors, the divergence in relative terms was much smaller than for the earlier years. In the same vein, the targets for money supply were exceeded by wide margins, especially in 1988 and 1990. This situation persisted in spite of the several mopping up operations of the Central Bank of Nigeria. The inability to keep to the money supply target or the targets for credit does not reflect the tight monetary posture of the Central Bank.

The major objective of the deregulating interest rate determination is to ensure efficient allocation of resources. Implicit in this assumption is the expectation that real interest rates would be positive. In this respect, performance has been partially successful as interest rates for 1988 and 1989 were decisively negative. However, the nominal interest rates rose significantly during the period.

The determination of the exchange rate by market forces was to remove or minimize price distortions, thereby effecting a more efficient allocation of resources. The exchange rate has depreciated the naira value significantly since the introduction of SAP. The act appears to have been overplayed since the naira is now generally believed to be grossly under-valued. International price comparison which the massive depreciation was to correct has thus become ridiculous in terms of the worth of naira.

Though the investment climate has improved tremendously with the liberalization of procedure for the repatriation of profit and new industrial policy, both domestic and foreign investment on new projects have been marginal. Replacement costs, let alone new investment, have become virtually impossible, given the new interest and exchange rate structures. Restoration of confidence of foreign investors in the Nigerian economy does not appear to have been achieved.

Corporate planning has also become increasingly difficult because of the relative volatility of both the interest and the exchange rate.

A target budget deficit not exceeding 3.5% of GDP was set under SAP. However, the figure has not been achieved since the inception of the program. This is in spite of massive reductions in subsidies for social services and for consumer products such as petroleum products as well as fertilizer.

Foreign debts have increased from between US$22 billion and US$25 billion in 1985 to about US$33 billion in 1990, reflecting the effect of re-scheduling and further draw-down on existing loans.

Reliance on market forces is the major thrust of SAP. The deregulation of the interest rate, the introduction of the foreign exchange market, and the reduction of the effective rate of protection in the industrial sector are in line with this thrust. However, the market itself imposes a constraint on the development process. Credit and foreign exchange are more readily available to large-scale enterprises and borrowers because short-term economic activities are favored over activities with long gestation period. In short, the market is bedeviled with greater imperfections than before and this impedes growth and development.

NIGERIA'S POLICY OBJECTIVES OF THE 1990S

The government had introduced in 1989 a new Industrial Policy Document which will chart a new course for the industrial sector in the 1990s. The document brought together various measures adopted under SAP which have an impact on industrial development. It noted that a major problem of the industrial sector was inadequate supply of imported inputs and spare parts, resulting in gross under-utilization of installed capacity. Other problems plaguing the sector, according to the document include geographical concentration, high production costs, low value-added, high-import content, and low level of foreign investment.

The policy focus will therefore be to:

- encourage the accelerated development and use of local raw materials and intermediate inputs rather than the dependence on imports:
- develop and utilize technology;
- maximize the growth in value-added of manufacturing production;
- promote export oriented industries;
- generate employment through the encouragement of private-sector small and medium scale industries;
- remove bottlenecks and constraints that hamper industrial development; and
- liberalize controls to facilitate greater indigenous and foreign investment.

To achieve these objectives, government would continue to put in place:

- a realistic and flexible exchange rate policy that will reflect the scarce nature of foreign exchange and therefore ensure its efficient allocation;
- tariff structures that will ensure effective protection of industries; and
- incentives that would improve the investment climate

The role of the public sector has been streamlined as follows:

- Encouraging increased private sector by privatizing government holdings in existing industrial enterprises
- Playing a catalytic role in establishing new core industries
- Providing and improving infrastructural facilities
- Improving the regulatory environment
- Improving the investment climate prevailing in the country
- Establishing a set of industrial priorities
- Harmonizing industrial policies at Federal, State, and Local Government levels

EXPECTED ROLES OF THE PRIVATE SECTOR

The failure of past industrial development policies has, since the mid-1980s, led to the search for alternative strategies. FMI therefore in 1988 adopted a new blueprint for industrial development which gave prominence to the role of the private sector. The FMI subsequently sought the assistance of UNIDO to put in place a new management approach to industrial development under an industrial master plan (IMP) otherwise referred to as strategic management of industrial development. In this new approach, the industrial landscape is seen as a system, describing a network of relationships, among various actors. The industrial system, using criteria such as resource base, production process, and/or market orientation, is sub-divided into subsystems for operational purposes. The objective is to identify key subsystems whose development would provide a catalytic role in the development of the whole system.

In giving effect to this new management approach, the FMI in August 1988 established the National Committee of Industrial Development (NCID) to, among other things, collaborate with UNIDO in executing this new management approach.

STRATEGIC MANAGEMENT OF INDUSTRIAL DEVELOPMENT

The IMP or Strategic Management of Industrial Development is predicated on the need to organize a network of actors around an industrial activity with a view to having a comprehensive and perspective view of the investment problems in that particular line of industrial activity. This network of actors is referred to as the Strategic Consultative Group (SCG). For a given industrial activity around which an SCG is organized, membership is drawn from manufacturers, raw material suppliers, transporters, policy makers, providers of infrastructural support services as well as distributors of the industrial goods/services.

Each SCG is expected to examine the problems of investment in its subsystem against the background of the existing policy environment, technology, structures of production, and market potentials. The SCGs are then expected to evolve a workable investment program toward restructuring, re-orienting, or developing new product lines which would ensure greater efficiency and competitiveness of the subsystem in both the domestic and international markets.

It should be noted that in the past, a major weakness of previous industrial and indeed overall development plans was that projects were planned, sectoral policies were articulated, and incentives were put in place without adequate consideration of the effect of one project on another, or the disincentive effects of an incentive policy for activities that are not covered by the particular incentive structure. In addition, policies and programs—due to political instability and/or political expediency—were subject to rapid changes. Worse still, many of the policies and programs of government were put in place under a near-total absence of current facts and figures on the Nigerian economy, and without sustained interaction and consultation with target beneficiaries of policies.

The macro-economic implications of the above situation are rather too obvious to warrant much elaboration. We only need to note the off-setting tendencies of inconsistent policies and incentives on the economy as a whole, as well as the mutual suspicion between public and private sector operators.

The Strategic Management of Industrial Development otherwise known as the industrial master plan (IMP) seeks to minimize the problems of policy and program inconsistencies in the development of a nation's industry. This, the IMP seeks to achieve through a systematic policy articulation that clearly spells out the inter- and intra-sectoral relationships within industry and between industry and other sectors of the economy including, of course, the external sector. It also seeks to modernize and rationalize existing industries in order to enhance their efficiency and competitiveness, as well as set up the institutional framework for the strategic management of the industrial sector. Strategic Management facilitates a clearer grasp of the inter-sectoral resource flows and their implications. It encourages and promotes constant interaction and consultation between various actors in the economic scene, especially between public sector policy makers and private sector operators. The interactive and consultative processes lead to mutual perception of problems, objectives, and strategies and thus minimize conflicts that usually arise in the legitimate pursuit of various group interests. In a way the IMP strategy seeks to temper the mood of society with objective realities of market forces and vice versa. It greatly facilitates information flow thus helping to reduce transactions costs.

In short, IMP can be described as a framework which provides for a dynamic and a regulated flow of investment funds and therefore a regulated industrial development in an environment of deregulation. The regulation is exercised through a system of industrial incentives and institutional support which then allow domestic and foreign entrepreneurs to interact, to invest, and to operate in areas that they calculate will maximize their returns. A careful and sustained monitoring of the responses of the private entrepreneurs and the associated resources flow to policy signals in turn enables government to adjust the incentive structure and institutional framework in an effective manner. In a way, therefore, the IMP is an attempt to promote medium- to long-term investment through the provision of clear, articulate, and "negotiate" industrial policy. The negotiation is undertaken through organized consultations between the private and public sectors in the course of the plan formulation and its implementation. The key objectives of the industrial master plan can be summarized as follows:

- Achievement of an orderly and coordinated industrial development
- Provision of a stable environment and a negotiated incentive structure necessary for medium- to long-term investment by the private sector
- Domestication of the industrial process through the promotion of local sourcing of industrial raw materials and the development of domestic technological capability
- Enhancement of economic efficiency so as to improve Nigeria's international competitiveness in the export of manufactured goods and thereby increase the country's export earnings and reduce the burden of debt
- Full development and exploitation of Nigeria's potentially large domestic market
- Provision of a flexible industrial base capable of quick but non-disruptive adjustments to national and international shocks
- Diversification and restructuring of the industrial base—diversification strengthens the economy by providing a cushion of shocks; that is, slacks in some economic activities are offset by positive developments in other areas, while restructuring seeks, in particular, to correct the imbalance between investment in consumer and capital goods production
- Maximization of the benefits of industrial development through well-designed inter-sectoral linkages; the benefits being, in particular, employment, outputs, material welfare, and some degree of national prestige

Over the short term, therefore, the IMP seeks to correct the shortcomings of the structure of industry and identify problems which must be resolved over the medium- and long-term periods so that the long-term objectives are attainable. Industrial priorities and the sequence of industrial development are also specified over this period. For example, the over-dependence on imported raw material and poor infrastructures are short-term problems, effective domestication of industrial structure will be pursued, that is, domestication of technology and industrial skills and the realization of the full benefits of industrial development, guided by the vision of industrial development already specified in the short-term.

STRATEGIES FOR IMPLEMENTATION

In outlining the modalities for the preparation and implementation of the industrial master plan (IMP) it should be noted that the focus on the private sector as the moving force in industrial development, and the implied decrease in the role of the government in direct productive investment, increase rather than diminish the complex role of the government in the management of the economy. It will require enhanced public sector professional and technical skills as well as developed institutional and political frameworks.

The implementation of the IMP involves three main phases. The first phase involves the formulation of the strategic guidelines for the IMP and the setting up of the institutional framework for the strategic management of the industrial sector. Phase two involves the formulation of strategic investment and action programs for industrial development, while phase three is to mobilize resources to implement and to monitor the investment and action programs.

In other words, the three interrelated phases imply first identifying and analyzing the problems of the industrial sector. Based on the analysis of the problems identified, design the appropriate investment strategies with given levels of investment related to time-specific growth targets. In order to ensure that investments are undertaken and targets are met, resources need to be mobilized and a strategic management and monitoring of the investment program put in place. The constant monitoring of the responses of private sector investments to given incentives provides the guide to appropriate adjustments when necessary. The unique advantages to these three interrelated phases of the exercise can be summarized as follows:

- The prior study of the chosen areas of focus and the consensus between all actors regarding an enhanced position of knowledge and facts of the economy and a mutual understanding of each other's roles
- The formulation of investment plans based on an earlier problem identification and analysis makes for a clearer articulation of financing gaps. This in turn provides an objective basis to both investors and financiers to reach a common ground on terms and conditions of lending
- The institutional framework for monitoring the investment programs, namely the Strategic Consultative Group (SCG) ensures timely and mutually agreed modifications to policies and programs in a non-disruptive manner

It is hoped that the new Strategic Management of Industrial Development would become a working reality that will put industrial development in Nigeria on a firmer footing through active participation and indeed leadership of the private sector.

PROBLEMS OF IMPLEMENTATION

From the planning experience discussed above, the following four major drawbacks in Nigeria's planning systems are apparent.

- Framework of objectives, priorities, and strategy are not often properly grounded in social reality. The planners do not show commitment to or identification with the vast majority of Nigerians: this in turn means that there is little or no attention paid to the plan.
- The macro-economic framework used is unduly simplistic. It is often based on anticipated money flows, and the aggregate plan is derived from the basic

model in which public sector leads and private sector follows. There is no direct link between projects and those that will implement them.

• Project selection process is so poor and disjointed that the exercise often appears as merely filling in the plan, and not what is expected to be implemented on the ground.

• The link between plan formulation and plan implementation leads to a situation where the plan merely lists the objectives at which policy measures will be directed without providing details on the implied relationships between policy instruments and the various target variables.

Specifically, the plans showed that some of the various structural weaknesses of Nigeria's industrial sector are as follows:

• Heavy import dependence because of the low stage of development of intermediate goods and capital goods industries
• Inadequate linkages among the various industrial sub-sectors and between industry and other sectors such as agriculture, mining, and construction
• Technological dependence because of the rudimentary engineering industries

NIGERIA'S STRATEGIES OF THE 1960S

During the colonial era, Nigeria did not have any explicit policy on industrialization. The colonial administration regarded the colonial territory as a source of supply of raw materials and one to serve as market for industries in Britain. It only embarked on minimum processing of natural resources to avoid carrying unnecessary financial burden. This was why the Ten-Year Plan of Development and Welfare (1946–1955) did not envisage any real industrialization program.

By 1950, there were only 20–30 factories that could pass for industries. These included palm oil milling, palm kernel crushing, groundnut crushing, cotton ginning, leather tanning, sawmilling, beer brewing, oil seed milling, among others. All these accounted for no more than 2.7% of the GDP. Therefore, one can safely say that Nigeria's industrialization started with primary processing of local raw materials before export.

Between 1955 and 1966, more local resource-based industries to supply the local market were developed. These included soap making, soft drinks, tobacco/cigarettes, bakeries, confectioneries, etc. This approach followed the classic theory of industrialization which advocates that newly industrializing countries should start by studying their imports and try to substitute the products by local manufacturing, beginning from simple consumer goods to more complex consumer durables. This gave birth to the import substitution or resource-based strategy which was adopted in the first national development plan (1962–1968). Many of the industries were small-scale industries as revealed by the industrial survey of 1963 which reported only 160 establishments employing 100 persons or more.

The import substitution strategy favored consumer goods industries which relied heavily on imported machinery and raw materials. The imported items were brought in with minimal or no duty, while the industries themselves were protected by high tariffs as "infant industries."

During the first decade of independence, the Nigerian government did not have much capital, and therefore investment in industry was left largely to foreign-dominated private sector. However, a number of industrial ventures were started under the sponsorship of some federal agencies and the regional development boards. Some of these industries include cement plants, breweries, cocoa processing, steel rolling mills, etc. In addition, market forces, fiscal incentives, and intensified program of import substitution motivated the establishment of medium- and large-scale manufacturing plants besides agricultural processing industries. The manufacturing sector growth, though still small in relation to GDP, had an annual growth rate of 15%–20% between 1960 and 1965.

The import substitution strategy has the following structural weaknesses:

- Over reliance on imported inputs
- A bias for the production of consumer goods
- Predominately internal market oriented for its output
- Over-dependence on protection from external competition, which bred inefficiency and market distortions
- Capital-intensive method of production
- Over-concentration on secondary stage processing with little or no internal linkages in the economy

The increasing pressures brought about by the implementation of the first national development plan, as well as the increasing social and political burdens of the newly independent administration, led to a call for foreign investment. Unfortunately, the response of foreign investors to this attempt to assert economic independence was poor and nonchalant as evident in the 1964 national budget speech which declared:

The classical era has come and gone: laissez faire is no more: the model of society has become highly agitated. It is obvious that economic progress cannot be divorced from political interference and national sentiment.

In spite of Nigeria's many natural attractions to investors such as extensive market, natural resources, and enormous amount of entrepreneurial talent, factors had tended to discourage investment in the manufacturing sector. The most important of these factors included uncertainty about the policy environment, a restrictive and cumbersome regulatory framework, and inadequate incentives.

NIGERIA'S STRATEGIES OF THE 1970s

In the 1970s, the growth of manufacturing was fairly rapid, averaging about 13% during the period. Consumer goods accounted for over 70% of total manufacturing value added (MVA), while the balance of less than 30% was accounted for by the intermediate and capital goods. The decade saw the emergence of the public sector as the leader in industrial investment and development, hence a singular emphasis on public sector planning with little or no attention paid to the private sector.

The public sector investments in the capital goods industry arose out of the need to shift the emphasis of the economy from outward to inward looking, that is, internal reliance for bother capital goods and overall economic development.

Some of the disadvantages of the internal market-oriented import-substitution approach are:

- it encourages production of consumer goods with limited linkage effects within the economy;
- it perpetuates, through foreign private industrial investment and the gains to investors, the national disadvantage of international division of labor;
- there is negligible advancement in industrial technology;
- it promotes foreign domination;
- an unorganized industrial development (a stagnating and small industrial sector);
- the high protective tariffs that usually support the strategy have had unfavorable effect on the structure of the manufacturing industry as it tends to weaken the incentive to make good quality products and to improve productivity; and
- it gives weak linkage effects which is responsible for the drift from the rural to the urban areas.

The implications of some of the above disadvantages include:

- lack of capacity to produce capital goods;
- low level of acquired technical skill; and
- perpetual dependence on external sources for the solution of Nigeria's basic problems.

Some of the advantages of the strategy of import substitution are:

- impressive growth in gross output, value added, and employment;
- dominant source of growth, accounting for about 80 percent of the growth in the output of manufacturing industries.

The import substitution strategy of industrialization was continued during the second national development plan period. However, because of large oil revenues received in the early 1970s, the country added a strategy of public sector-led industrialization. These two planks in our industrialization effort had their advantages and disadvantages, although later developments in the economy tilted the balance in favor of the advantages, thus showing up the structural weaknesses of the nation's industrial sector.

However, at the beginning of the 1970s, the manufacturing sector was still being dominated by foreigners. This development which was considered unsatisfactory led to the formulation of indigenization policy as enunciated in the NEPD of 1972.

THE NIGERIAN ENTERPRISES PROMOTION BOARD (NEPB)

The Nigerian Enterprises Promotion Decree was promulgated in 1972 (amended 1977) to involve Nigerians in the ownership, control, and management of certain enterprises. The degree was designed to:

- create opportunities for Nigerian indigenous businessmen;
- raise the proportion of indigenous ownership of industrial investments;
- maximize local retention of profits;
- raise the level of local production of intermediate goods;
- increase Nigerian participation in decision-making in the management of larger commercial and industrial establishments;
- advance and promote enterprises in which citizens of Nigeria shall participate fully and play a dominate role;
- advise the Minister of Industries on clearly defined policy guidelines for the promotion of Nigerian enterprises;
- advise the Minister on measures that would assist in ensuring the assumption of the control of the Nigerian economy by Nigerians in the shortest possible time; and
- determine any matter relating to business enterprises in Nigeria generally in respect of commerce and industry that may be referred to it in accordance with any direction of Minister, and to make such recommendations as many as necessary on those matters in such manners as may be directed by the Minister.

All enterprises were classified under three schedules, each of which indicates the maximum equity participation a foreigner could hold in the enterprises listed under it. The list under each schedule has been modified from time to time to reflect developmental priorities.

During the period of its existence, the NEPB played a critical role as a promoter of indigenous enterprise. However, one of the major weaknesses of the Decree was that the foreigners quickly realigned their enterprises to seek for exemption. Therefore, from the outset, the Board faced the problems of:

- identifying the enterprises affected by the Decree;
- recruiting adequate number of qualified staff to man the various organs of the system;
- finding necessary finance to acquire the enterprises affected; and
- training the required manpower to replace the alien owner-managers in Schedule I enterprises.

As a result of the above shortcomings, the 1972 Decree was reviewed and the 1977 Nigerian Enterprises Promotion Decree was promulgated to correct them. The features of the new Decree designed to increase its effectiveness were as follows:

- Every enterprise was affected.
- State promotion committees were created to supplement the activities of the Board in their respective states.
- Zonal offices were opened to increase operational effectiveness of the Inspectorate Department.
- The NBCI and NIDB as well as other banks and financial institutions were specifically mandated to give loans to prospective Nigerian investors.

During the 1980s, it was observed that total investment as a share of GDP had fallen due to several factors ranging from inadequate foreign capital inflow to low levels of internal savings. This situation led Government to review the investment environment particularly in the light of the fact that Nigerian entrepreneurs have come of age, and are able to hold their ground in various types of enterprises. The review was also necessitated by the fact that demands of SAP imposed the spirit of competition and efficiency in production and quality of goods and prices acceptable to consumers.

Therefore, in order to encourage foreign capital inflow, the government amended the NEPD of 1977. With the amendment, there now exists only one list of scheduled enterprises exclusively reserved for Nigerians for the purpose of 100% equity ownership. All other businesses not contained in the schedule are now open for 100% Nigerian for foreign participation except in the areas of banking, insurance, petroleum prospecting, and mining where the previous arrangements still subsist. This new ownership structure applies to new investments only.

Financing and incentives, trade information, trade facilities, export publicity, and training. Despite the existence of the NEPC, the following structural problems militated against export promotion:

- Poor product quality
- Uncompetitive prices
- Inadequate transportation
- Supply bottlenecks

Therefore, in order to further encourage the export of agricultural and manufactured products, the Federal Government promulgated the Export Promotion Decree (No. 18) of 1986. The decree provided for the following incentives:

- Retention of export proceeds in foreign currency in a "domiciliary" bank account in Nigeria. Such money may be used to pay for specified activities
- Abolition of export license requirements for the exportation of manufactured or processed goods

- Export Credit Guarantee and Insurance Scheme
- Scrapping of Commodities Boards which by their monopolistic nature hindered the operations of free-market forces
- Export Development and Expansion Funds to assist exporters to cushion some of their initial expenses and to boost their future consignments
- Export Adjustment Scheme Fund which is a form of subsidy on cost of production;
- Re-discount of Short Term Bills for Export
- Additional capital allowance where applicable
- Tax relief on interest on capital invested in export-oriented industries
- Export-Free Zones in selected states of the federation

INDUSTRIAL TRAINING FUND (ITF)

The ITF, established by Decree No. 47 of 1971, is the body responsible for promoting and encouraging the acquisition of skills in industry and commerce. It is expected to continue to generate indigenous trained manpower to meet the needs of the economy. To this end, the fund facilitates training of persons employed in industry and commerce, approves courses, and appraises facilities provided by other bodies. It also assists individual persons or corporate organizations in finding facilities for training for employment in industry and commerce and conducts or assists others to conduct research into any matter relating to training in industry.

STRATEGIES OF THE 1980s

At a meeting held in 1980, the African Heads of State adopted the Lagos Plan of Action designed to promote the twin goals of promoting self-reliance in food production and self-sustaining industrialization. In order to facilitate the industrial component, the Heads of State declared the 1980s the Industrial Development Decade for Africa (IDD). The IDDA programs operationalized the concept of self-sustaining industrialization to include:

- identifying and establishing core industries such as iron and steel, petrochemical fertilizer, and pulp and paper projects;
- reassessing industrial strategies toward a local resource-based industrialization; and
- creating internal engines of growth.
  The objectives of the Lagos Plan of Action and IDDA are therefore to:
- reduce dependence on external demand stimuli;
- reduce dependence on external factor input or other supply; and
- internalize employment and income multiplier effects of investment.

We have noted earlier that the industrialization policies of the 1960s and 1970s have been mainly geared toward the promotion of import substitution and the manufacture of consumer goods. Although import substitution is not fundamentally bad, it should not have been predicated upon the importation of raw materials and components, and should not, as is often the case, be a mere assembly operation which contributes neither to the upgrading of indigenous resources nor to the development of technological potentials. The strategy also prolonged the dependence on external sources while the creation of capital-intensive import-substitution industries distorted cost structures. For emphasis, it is necessary to again restate those other recurring structural weaknesses of the industrial sector that include the following:

- Lack of base from which Nigeria could launch into a self-sustaining industrial growth since there was, virtually, no engineering industries or well-developed capital goods sector, and was, therefore, vulnerable to external shocks
- Little or no real linkage industries or inter-sectional transactions
- Lack of requisite stick of human capital with the necessary technological skills for an industrial take-off

POLICY PACKAGE

In order to revitalize the manufacturing sector and to accelerate its pace of growth and development, government adopted an industrial policy package which was intended to:

- increase industrial value-added;
- diversify the industrial base by supplying basic and intermediate inputs;
- generate productive employment and raise productivity;
- increase the technological capacity;
- increase private sector participation in the economy; and
- disperse industries evenly across the country.

The objectives of the fourth national development plan were therefore designed to tackle the issues mentioned above. However, the plan was more or less stillborn as the economy was then beset with problems which had their origin largely in the worldwide recession, and partly in poor economic management. By 1985, the economy was virtually on the verge of collapse.

Capacity utilization of most industries was below 20% owing to lack of foreign exchange for raw materials and spare parts, and inflation had attained an intolerable level. It therefore became clear that a structural reform was inevitable. It was in this context that the SAP was introduced in July 1986.

NIGERIA'S INDUSTRIAL POLICY OF 1980s

A revised national industrial policy was launched in 1980 with the following objectives:

- Maximization of local value-added
- Promotion of research and development

- Employment generation
- Promotion of export-oriented industries
- Industrial dispersal

The strategies for achieving the objectives were as follows:

- Full recognition of private enterprise and initiative as the responsibility of the state, for the welfare of every citizen
- Provision of adequate incentives to industries

Then industrial policy could not be fully implemented as a result of the oil crisis in 1981 which led to a dwindling of Nigeria's foreign reserve.

When the military took over in 1984 following the economic crisis of 1982–1983, the administration looked inward and took some decisive steps to discipline the populace by curbing their extravagant lifestyles that depended heavily on imports. The government refused to reopen negotiations with IMF, and preferred austerity and prudent economic management to fight the problem.

When in Babangida administration assumed office in August 1985, it introduced a period of economic emergency which was to last for 15 months. The Decree granted the present wide powers to deal with the problems of the economy. The government also sought and used to introduce an economic recovery program which aimed at altering and realigning aggregate domestic expenditure and production patterns so as to minimize dependence on imports, enhance non-oil export base, as well as bring back the economy on the path of stead and balanced growth.

The 1986 budget gave the main outline for the structural adjustment program which was published later that year.

The problems of industry were specifically analyzed and provided for in the SAP document, It noted that given the structure of Nigeria's industries, its major problem was inadequate supply of imported inputs and spare parts, resulting in gross under-utilization of installed capacity.

Government therefore came to the conclusion that local sourcing would be the long-term solution, since foreign exchange will never be adequate. The SAP document provided specifically for the industrial sector as follows:

- Encouraging the accelerated development and use of local raw materials and intermediate inputs rather than depend on imported ones
- Development and utilization of local technology
- Maximizing the growth in value-added of manufacturing production;
- Promoting export-oriented industries
- Generating employment through the encouragement of private sector small- and medium-scale industries
- Removing bottlenecks and constraints that hamper industrial development including infrastructural, manpower, and administrative deficiencies
- Liberalizing financial controls to facilitate greater indigenous and foreign investment

THE STRUCTURAL ADJUSTMENT PROGRAM

The major objectives of the SAP are:

- to restructure and diversify the productive base of the economy in order to reduce dependence on the oil sector and on imports;
- to achieve fiscal and balance of payments viability over the medium term; and
- to lay the basis for a sustainable, non-inflationary growth over the medium and long term. For the industrial sector, the strategy under the program aimed at:
- encouraging the accelerated development and use of local raw materials and intermediate inputs rather than depend on imported ones;
- development and utilization of local technology;
- maximizing the growth in value-added of manufacturing production;
- promoting export-oriented industries;
- generating employment through the encouragement of private sector small- and medium-scale industries;
- removing bottlenecks and constraints that hamper industrial development including infrastructural, manpower, and administrative deficiencies; and
- liberalizing controls to facilitate greater indigenous and foreign investment.

In order to achieve these objectives, the government put in place a number of measures which have materially changed the macro-economic framework in which industries have to operate. Major aspects of the new environment include:

- emphasis on private sector-led growth;
- a measure of deregulation of erstwhile controls on economic activities;
- liberalization of trade but providing minimum protection to strategic industries;
- freer access to foreign exchange market where the exchange rate for naira is determined by the interplay of market forces;
- privatization and commercialization of some government investments; and
- a more open, competitive economy;
- abolition of commodity boards;
- abolition of the Import Licensing Scheme; and
- removal or substantial reduction of various government subsidies by means of commercialization of some utility services such as electricity supply and telecommunications services.

IMPACT OF SAP ON MANUFACTURING ACTIVITIES

It is necessary to recapitulate the economic climate under which SAP was introduced in July 1986, a gloomy background of mounting external debt, unhealthy investment, and the failure of the regime of stringent trade and exchange controls which had been pursued in the previous two decades.

As a consequence of SAP, many large enterprises in agro-industry diversified into the growing and processing of primary agricultural produce for further processing into raw materials for their plants.

Also, due to improved access to foreign exchange and consequently to imported inputs, capacity utilization of the industrial sector increased on the average from 30% in 1986 to 37% in 1987 and 40% in 1988. Very high-capacity utilization was experienced in domestic resource-intensive activities such as textiles, furniture, rubber, and non-metallic mineral products, while low capacity utilization was found among import-oriented activities with relatively low domestic value added such as car assembly, electrical and electronic plants, basic chemicals and pharmaceuticals.

In conclusion, the decade of the 1980s witnessed a deep crisis in the nation's industrial development. The crisis is evidenced by:

- a steady decline in capacity utilization;
- deterioration of the tools of production and decay of capital assets in some cases;
- low rate of investment in industry and even in some cases, disinvestment; and
- high cost, low quality, and hence uncompetitive products.

In the 1980s, Nigeria's industrial landscape consisted of industrial ventures that were ridiculously dependent on imports, with crushing effect on balance of payments as well as the stock of scarce foreign exchange. These and more were the trend SAP sought to reverse.

NIGERIA'S STRATEGIES OF THE 1990S

There was a deep crisis in our industrial development during the 1980s. As a result, the aims and objectives of the "first" Industrial Development Decade for Africa (IDDA) were largely unfulfilled, and they have been re-stated for the "second" decade of the 1990s. Some of the lessons learned during the 1980s included:

- the need to reappraise policies and strategies;
- the need to reduce dependence on external factors for our development, and as a corollary; and
- the need to create internal engines of growth so as to internalize the multiplier effect of investment.

Therefore, the experience of the 1980s and of past development planning efforts convinced the government of the need for a change in the development planning strategy for the country. A plan for the industrial sector would define the sector's forward and backward linkages with other sectors of the economy such as agriculture, transportation, construction, communication, mining, and energy. This led government to decide on the mechanism of a Perspective Plan and

Three-Year-Rolling Plans, so as to put in place "plans that are subject to periodic reviews rather than fixed plans." The plan is expected to forge a closer link with the annual budget.

Perspective Plan

Nigeria has adopted a 15–20-year perspective plan for the purpose of taking a long-term structural view of the economy. The plan was launched in 1990 with the following objectives:

- To specify long-term socio-economic development goals and targets
- To articulate the developmental options open to the nation over the longer term
- To identify possible bottlenecks and problems in the way of achieving these developmental goals and targets
- To articulate policies and strategies for eliminating the bottlenecks and effectively pursuing the long-term goals and targets
- To specify sectoral investment priorities; and, importantly,
- To ensure consistency in the management and trajectory of the economy over the long term

The key objectives of the plan are:

- attainment of higher levels of self-sufficiency in the production of food and other raw materials;
- laying of solid foundation for a self-reliant industrial development as a key of self-sustaining dynamic and non-inflationary growth, and promoting industrial peace and harmony;
- creating ample employment opportunities as a means of containing unemployment problems; and
- enhancing the level of socio-political awareness of the people and further strengthening the base for a market-oriented economy and mitigating the adverse impact of the economic downturn and the adjustment process on the most affected groups.

Rolling Plans

The strategy of the rolling plan is to consolidate the achievements that have been made so far from the SAP. A major aspect of the SAP is the creation of the appropriate policy environment that would promote the growth of the direct production sectors of the economy, that is, agricultural and industrial sectors through effective mobilization of available development resources, promotion of inflow of foreign investment, and efficiency in resource allocation through a pricing system that responds appropriately to market signals.

The first of the Three-Year-Rolling Plans, 1990–1992, echoed the Lagos Plan of Action and the SAP objectives. It stated inter alia that the "objectives of self-sufficiency in the production of food and agro-based raw materials is in line with the current efforts to expand the productive base of the economy."

The Rolling Plan 1990–1992 has identified the following major constraints on the industrial sector:

- Shortage of industrial raw materials and other inputs
- Infrastructural constraints
- Inadequate linkage among industrial sub-sectors
- Administrative and institutional constraints

Whereas the rolling plan and the perspective plan give indications of qualitative and quantitative changes envisaged for the economy, it was still considered necessary to have sectoral master plans which will contain details of how those changes are to be programmed and implemented. As a result, government decided to embark on an industrial master plan.

Nigeria's Industrial Master Plan

As stated in the First National Rolling Plan (1990–1992), the industrial master plan is "aimed at promoting the development of an efficient industrial system through the determination and definition of all the functional aspects of an industrial system, and the preparation of an action plan to achieve established objectives and targets."

In this case, an industrial master plan (IMP) is merely a tool for the strategic management of industrial development. It provides the basis for guiding investment into productive facilities, support services, and training.

The developmental issues to be addressed by the IMP would include:

- local resources as factors input;
- technological issues;
- manpower development for industrialization; and
- physical and institutional infrastructure for industrial development.

The strategy of industrial master plan has now been translated into what is called Strategic Management of Industrial Development (SMID). Like the master plan, the SMID is a tool for management. SMID sets in motion the various processes, institutions, and stages leading to the strategic choices of a path for the industrial development. It calls for re-orientation on the proper role of the state in industrial development, especially in the context of a market-oriented, open competitive system implied by SAP.

In adopting the new strategy, the government has started to withdraw from certain areas of productive investment, through its privatization and commercialization program.

Also, through various policies, the government is installing a more open and de-regulated market economy.

THE MOVE TOWARD PRIVATIZATION

The strategy of the privatization/commercialization of public enterprises was adopted in July 1989 by the Federal Government.

Privatization as a public policy was toyed with in the early 1980s. The Onosode Commission (1982) and a study group set up by the Buhari Administration in 1984 recommended the privatization of certain parastatals. However, the issue did not evoke much public debate until president Babangida in his 1986 Budget Speech announced government's decision to adopt a privatization policy. Subsequently, in July 1988, the Federal Government promulgated the Privatization and Commercialization Decree No. 25 of 1988, which provided the legal framework for the implementation of the program within the context of the ongoing restructuring of the Nigerian economy. The Decree empowers the Technical Committee on Privatization and Commercialization (TCPC) to implement the program.

The basic objectives of the program are to:

- restructure and rationalize the public sector in order to lessen the dominance of unproductive investment in the sector;
- re-orientate the public enterprises for privatization and commercialization toward a new horizon of performance improvement, viability, and overall efficiency;
- ensure positive returns on public sector investment in commercialized enterprises;
- check the present absolute dependence on the treasury for funding by otherwise commercially oriented parastatals and so encourage their approach to the Nigerian Capital Market; and
- initiate the process of gradual handing over to the private sector such public enterprises which, by their nature and type of operation, are best performed by the private sector.

The enterprises affected are spelled out in Schedule I (privatization) and Schedule II (commercialization) of the Decree. The listings are in the following four categories:

Category I 37 enterprises in which equity held by the Federal Government shall be partially privatized;

Category II 67 enterprises in which 100% of the equity held by the Federal Government shall be fully privatized;

Category III 14 enterprises billed for partial commercialization, and

Category IV 11 enterprises to be fully commercialized.

In all, a total of 129 enterprises are affected by the exercise. Some of the apprehensions of the policy include the fact that:

- the exercise could be compromised by interest groups and political considerations thereby replacing public monopoly with private monopolies;

- workers in the privatized companies are likely to suffer certain disadvantages, because most of them are unlikely to be able to buy equity shares, their remuneration could be reduced and their job security could be jeopardized; and
- it could open the floodgate for foreign investment, thus eroding the gains of the indigenization exercise of the 1970s.

Other issues which have been raised on the program relate to:

- control, especially of enterprises in which government relinquishes its entire shareholding (i.e., 100% privatization) without the presence of a core group of shareholders with adequate knowledge or experience of the particular business; and
- ensuring adequate compensation for government investment by way of equity and loans having regard to the age of such public sector investments.

NIGERIA INDUSTRIAL POLICY OF 1989

The new industrial policy was put in place in 1989. The policy document brings together various measures that had been adopted under SAP which have an impact on industrial development. It accepts the fact that the realization of the objectives of accelerated industrial development hinges on the response of the private sector to the new set of policies.

The role of the public sector has been streamlined as follows:

- Encouraging private sector participation by privatizing government holdings in industry
- Playing a catalytic role in establishing new core industries
- Improving the regulatory environment
- Improving the investment climate prevailing in the country
- Establishing a clear set of industrial priorities
- Harmonizing industrial policies at federal, state, and local levels of government

To achieve the above objectives, government is reforming the institutional framework that regulates industrial investment in the country. Some of these institutions are discussed below.

INDUSTRIAL DEVELOPMENT COORDINATING COMMITTEE

One of the problems which used to militate against foreign investment in Nigeria was the restrictive and cumbersome regulatory environment under which the economy operated. As a result, government in formulating its new industrial policy established an institutional arrangement for investment promotion through Decree No. 36 of 1988 which established the Industrial Development

Coordinating Committee (IDCC). It is an inter-ministerial committee to ensure speedy decision-making in matters concerning new investments in industry.

The committee was set up to improve the investment climate and attract investors by centralizing and simplifying procedures for issuing all pre-investment approvals. It is a one-stop agency for all pre-investment approvals. This approach removes duplication, delays, and frustration that tended to scare away genuine investors.

SMALL-SCALE INDUSTRIES

The present industrialization strategy is predicated on the development of local resource-based industries and small-scale industries.

- A close look at the structure of the manufacturing sector shows that a large percentage of the industries produce consumer goods which are heavily dependent on imported basic and intermediate raw materials. For example, at the 1987 FEM bidding, about 40% of the foreign exchange disbursement went to the importation of raw materials for manufacturing industries and about 35% to the importation of machinery and spare parts including completely knocked down (CKD) parts. As a result, government has decided to encourage industrial activities which rely heavily on resources which are abundantly available locally. This strategy will increase the multiplier effect of the industrialization process. Government will also encourage local production of intermediate goods and the fabrication of capital goods (of intermediate technology content) as well as spare parts.
- In the past few years, attention has been focused on the merits of small-scale industries. These include employment generation, greater utilization of local financial and material resources, promotion of indigenous technology, tremendous opportunities for sub-contracting by large-scale industries, and forward and backward linkages. Also, the initial capital required for investment is low, and the capital/labor and capital/output ratios are relatively low. Therefore, a rapid industrial development must be based on the development of small-scale industries. In the pursuance of this strategy, the local government areas in the country have been grouped into three zones based on criteria such as per capita industrial output, the degree/extent of social and economic infrastructures available, and the level of development of the labor market. The three zones are:
  - Zone 1: Industrially and economically developed local government areas
  - Zone 2: Less industrially and economically developed local government areas
  - Zone 3: Least industrially and economically developed local government areas

Government also intends to make all areas of the country attractive to new investors through a package of incentives, including a program of industrial layouts and craft villages development. The government will also assist state governments with matching grants in the establishment of industrial estates for small-scale industries. Other activities include the ongoing Entrepreneurial Development Program (EDP), the Working-For Yourself Programs (WFYP) and the Training the Trainers Scheme. These programs show great promise of developing the corps of entrepreneurs needed for successful implementation of the small-scale industrialization strategy.

INDUSTRIAL INCENTIVES

These industrial development strategies are complemented by a number of incentives which have been put in place by the government to encourage both indigenous and foreign investors. The existing incentives can be classified as follows: fiscal, trade/exchange rate policies, and export promotion measures. Others include special assistance programs, the provision of infrastructural facilities, and extension services at industrial estates as well as manpower training. Some of these are:

(a) **Fiscal Incentives**

Fiscal incentives are meant to provide for deductions and allowances in the determination of taxes payable by manufacturing enterprises. The existing fiscal incentives are:

Pioneer Status:
The Income Relief Act of 1958 amended by Decree 22 of 1971 provides that public companies be granted specific tax holidays on corporate income. The objective of this incentive is to encourage industrialists to establish industries which the government considers critical to the overall development of the country. It is also intended to attract foreign investment to Nigeria for the purpose of promoting industrial expansion and the development of the country's natural resources. The scheme discriminates in favor of industries with bias for:

- export-oriented activities;
- labor-intensive processes;
- local raw material sourcing;
- the development of infrastructural facilities; and
- on-the-job training.

The relief covers a non-renewable period of 5 years for pioneer industries, and 7 years for those of them located in economically disadvantaged areas.

Tax Relief for Research and Development:
Industrial establishments are expected to undertake R&D activities for the improvement of their processes and products. To this end, up to 120% of expenses on R&D are tax deductible so long as such research activities are carried out within the country and are related to the activities of the company. There is a higher allowance of 140% for the development of local

raw materials. In addition, the total capital expenditure might be written off against profits where the R&D is on a long-term basis.

Companies Income Tax:
The Companies Income Tax Act 1979 with its amendments is meant to encourage potential and existing investors by reducing the corporate tax rate to 40% from 1987. It also modifies the capital allowance rates to reflect changes in rates.

Tax-Free Dividends:
Under the specified conditions laid down by government, an individual or company deriving dividends from any company effective from 1987, shall enjoy tax-free dividends for a period of 3 years. The tax-free period shall be 5 years if such companies are engaged in agricultural production, the processing of agricultural products, the production of petrochemicals, or the production of liquefied natural gas.

Tax Relief for Investments in Economically Disadvantaged Local Government Areas:
Investors in economically disadvantaged local government areas are entitled to the following tax incentives:

- seven years income tax concessions under the pioneer status scheme;
- special concessions by relevant State Governments; and
- additional 5% over and above the initial capital depreciation allowance under the Company Income Tax Act (Accelerated Capital Depreciation).

(b) **Trade/Exchange Rate Policies**
Government has put in place a new tariff regime which provides for a considerable degree of protection consistent with the industrial development objectives of developing economies at a similar state of development.

INSTITUTIONAL FRAMEWORK

In an effort to restructure the economy, there exist certain institutional mechanisms which have direct consequences for the catalytic role which government must play under the new dispensation. These institutions include:

Industrial Inspectorate Department (IID)
The Industrial Inspectorate Department was established by Decree 53 of 1970. The functions of the IID are:

- certifying the actual values of capital investments in buildings, machinery, and equipment of various companies;
- certifying the value of imported industrial machinery for the purpose of granting approved status to non-resident capital investment; and
- monitoring of the Comprehensive Import Supervision Scheme to ensure that the operations are in the spirit of the relevant agreement.

Standards Organization of Nigeria (SON):

The Standards Organization of Nigeria is charged with surveillance over the products of Nigerian Industries and over imported products to ensure that they meet national international standards.

SON was established to arrest the poor quality of local products which has been traced to the raw materials and the machinery/equipment as well as the absence of quality planning and control procedures.

Policy Analysis Department (PAD):

PAD undertakes policy analysis necessary for the evaluation and effectiveness of industrial policies. These include:

- assessing the extent of anti-export bias which is related to the trade and exchange rate regime;
- evaluating Nigeria's export incentives schemes;
- evaluating other prevailing export regulations and their administration;
- assessing the impact of measures in terms of government fiscal revenues, volume of import and export, and in the light of industrial growth in the short and medium terms; and
- developing fiscal investment incentives.

Industrial Development Coordinating Committee (IDCC):

Introduced in June 1986 as a result of under delays and other problems which created disincentives for many genuine investors, the IDCC now serves as a central agency where all the approvals required for foreign investment in Nigeria can be obtained. It replaced the multiplicity of approving bodies.

The objectives of establishing the IDCC are:

- to obviate delays in granting approvals for the establishment of new industries by creating one approval center which will replace the multiplicity of approving centers, which, in the past, had been responsible for unnecessary and avoidable costs to prospective investors before approvals are granted;
- to advise on policy review proposals as they relate to industrial development;
- to ensure adequate coordination and objectivity in the nation's industrial development efforts;
- to make recommendations on pertinent industrial policies including tariffs and various incentive measures aimed at enhancing steady industrial development; and
- to ensure that industrial location is consistent with government's environmental policies.

Investment Information and Promotion Centre:

This center provides information on procedural matters and industrial incentives. It also advises and guides prospective local and foreign investors on most aspects of their investment proposals.

Industrial Data Bank:

This bank is responsible for storing and retrieving data in order to provide information on existing industries in the various sub-sectors. Such information includes production capacities, capacity utilization and expansion plans, production costs, the extent of the market, price movements, and raw material availability in various parts of the country.

National Office for Technology Acquisition and Promotion (NOTAP):

NOTAP, initially known as the National Office of Industrial Property (NOIP) was established by Decree No. 70 of 1979, to facilitate the acquisition of technology in Nigeria. The major objective of the office is to ensure that the acquisition of technology brings about social and economic gains in the areas of rapid industrialization, the development of local technologies, and the exportation of same. It is also to monitor all registered agreements to ensure that the implementation process complies with laid-down terms as stipulated in the clauses of the contract. The information generated from the monitoring exercise is utilized by government in the formulation of new policies and modification of existing ones.

In order to achieve government objectives in the technological development of the country, the office provides information and advisory services to Nigerian entrepreneurs to enhance the development of their negotiations, skills, and expertise in the acquisition of technology. It has also directed the activities of the industrial sector toward adaptive research, effective training, and the maintenance culture in all aspects of technology.

One can say that NOTAP has been able to provide effective support and assistance to Nigerian enterprises in their efforts to acquire technology at reasonable terms and conditions for rapid industrialization. Also, the activities of NOTAP have led to further improvement in the negotiating skills of entrepreneurs, while interactions with entrepreneurs have proved advantageous in leading to a reduction of serious defects in technology agreements. If commendable initiatives in Nigeria, such as NOTAP's accomplishments, are to be institutionalized throughout the nation, they must be implemented through a structured systems approach, akin to what DEJI Systems Model advocates. This is how Nigeria can leverage her science, engineering, technology, and innovation (SETI) initiatives for sustainable development. Segment 3 highlights the IDC failures that occurred as a result of a lack of integration. DEJI Systems Model could have helped to avert the problems reported in the segment.

SEGMENT 3: INTEGRATION

AKINWALE'S INVESTIGATIVE REPORT ON NIGERIA'S IDCS

The accounts presented here are based on the investigative report by Akinwale (2019) on the gradual decline of Nigeria's IDCs. He reports that the 23 Industrial Development Centres (IDCs), designed and built to service small- and medium-scale enterprises (SMES) in Nigeria, are derelict and abandoned despite Federal Government's plan to use the facilities to boost small-scale local business in the country. He visited IDCs in Lagos, Port Harcourt, Abuja, and Kano and discovered the dilapidated condition of the multi-million infrastructure built across the country. His findings could have been predicted by the direct observations of this author in 1994, at which time he raised an alarm about the decline and neglect of the IDC infrastructures. The only two viable IDCs that he observed in 1994 were the ones in Owerri and Oshogbo. His official report to UNDP and the Nigeria Ministry of Industry documented to findings and the unsavory predictions.

According to Akinwale (2019), the Nigeria's Industrial Development Centres (IDCs) in Abuja, Lagos, Port Harcourt, and Kano have been taken over by miscreants, farmers and herders, and the supervisory agency, Small and Medium Scale Enterprises Development Agency of Nigeria (SMEDAN) is not unaware of this but could only do little or nothing; The ICIR can report.

Though the Federal Government made repeated promises to rehabilitate the moribund 23 industrial centers across the states after admitting in 2016 that they have been abandoned and dilapidated for too long. The 23 industrial centers established by past governments were to serve as a support system for small and medium-scale enterprises (SMEs) in the country but none of them functions currently. The first IDC was established in Owerri in 1965 by the former Eastern Nigeria government, Ministry of Trade and Industry, and was taken over in 1970 by the Federal Government including the one in Zaria, Northern Nigeria, which was established in 1969. The emergence of the centers followed the Nigerian government's yearning to strengthen SMEs in the country. The centers were established and located where the country has a comparative advantage of natural resources. Experts who carried out feasibility studies recommended that the government should concentrate on five areas, namely woodwork, metalwork, automobile repair, textiles, and leatherwork. When establishing the centers, the Federal Government spent huge funds providing workshops, machines, offices, and vehicles. Though the centers were previously under the Ministry of Industry, Trade and Investment, they were later handed over to the Small and Medium Enterprise Development Agency of Nigeria (SMEDAN) in 2011. However, due to years of neglect by the government, the centers are currently wasting away. This was what this author predicted in 1994. The administration of President Muhammadu Buhari launched an micro, small, and medium enterprises (MSME) clinics across the states to fast-track development of SMEs but did not articulate any action plan in the clinic for the revival of the supporting centers. MSME was set up by the federal government of Nigeria through the Central Bank of Nigeria. It was launched on 15 August 2013 with a share capital of N220 billion and basically tends to finance micro, small, and medium enterprises in the economy. Investigative visits to the centers in 2017 by the Senate Committee on Industry did not yield any results despite lamentations by members of the committee about the sorry state of the centers (Akinwale, 2019).

The IDCs were established to provide extension services to SMEs in such areas as project appraisal for loan application, training of entrepreneurs, managerial assistance, product development, production planning and control, as well as other extension services. In 2017 and 2018 budgets, SMEDAN allocated N600 million to rehabilitate 12 centers. The amount was spent to construct fences around some of the facilities and to complete other renovation works.

SMEDAN has spent N600million on the rehabilitation of the IDCs between 2017 and 2018. Some of the works include erection of perimeter fences around the centers. According to Ibrahim Kaula Mohammed, the Head Corporate Affairs Department at SMEDAN, the agency inherited the dilapidated centers from the Ministry of Industry, Trade and Investment. "When we inherited these IDCs, that's how almost all of them were, dilapidated," Mohammed was quoted as saying (Akinwale, 2019). On why the agency has failed to revive the centers and allowed them to deteriorate, Mohammed, said, "We didn't receive any kobo at this agency for IDCs in terms of budgetary allocation until 2017. Any improvement you see on any IDC is only from 2017 and 2018."

The agency, in 2017, fenced the premises of IDC Benin, Katsina, Ikorodu, Owerri, and Abuja. In Zaria, it did a general renovation of the 3-in-1 workshop. In fact, this author had a direct encounter with the staff of IDC Zaria in 1994, in which a reported was supply project was represented by only a large concrete slab, with no visible inlet or outlet openings for any water flow. This was reported as a possible case of fraud and corruption, but nothing was done following that report. At the IDC Idu, Abuja, the roof was renovated. In 2018, the agency bought complete automotive component of waste to wealth equipment in Idu, Zaria, and fenced Kano, Makurdi, Jos, and Abeokuta centers. Despite all this spending, the IDCs would be demolished any time soon. A viability study has shown that despite the rehabilitation, they will be replaced with what is now known as industrial clusters. Hopefully, this time around, some techniques such as DEJI Systems Model will be employed to forestall potential operational problems down the line. SMEDAN and Africa Development Bank (AfDB) carried out a study of all the IDCs, which comes in two stages: outline business case (OBC) and full business case (FBC). It was gathered that the OBC has been completed and submitted to the AfDB which has the final say to go ahead with the second stage, FBC. It is expected that management would engage auditors to examine the work.

> That is what is remaining and we have done procurement in that respect. An auditor has been selected. It is now for AFDB to agree with what we have done. Once AfDB is satisfied with all the processes, we would now sign a contract with the auditor,

one official reported during Akinwale's investigative interviews (Akinwale, 2019). For the second part of the AfDB case, the FBC is going to demolish all those workshops (even those rehabilitated) because that is what the consultant who carried out the study proposed. Based on the proposal, there would be a vertical cluster that would accommodate more SMEs. The spokesperson said the concept all over the world is an industrial cluster, a group of interrelated business in one place making use of a common facility and infrastructure. This approach, if implemented correctly, aligns with the definition of a system presented in Chapter 1 of this book:

"A system is a collection of <u>interrelated</u> elements, whose collective output exceed the sum of the individual outputs of the elements."

Six centers—Abuja, Lagos, Sokoto, Port Harcourt, Owerri, and Maiduguri are said to be in the pilot stage of the proposed industrial clusters. A spokesman said all the 23 IDCs are viable, according to the study. "But we cannot start with the whole 23, we have assigned the cluster type to each of the six. The selection is based on the competitive advantage of each of the states." Lagos will focus on Fast Moving Goods (FMGs), while Port Harcourt will focus on chemicals and oil products because of the availability of petroleum there.

In Nigeria, small-scale businesses constitute 85% of all firms operating in the economy Hassan and Olaniran (2011). Like in most other developing countries, small-scale businesses employ the largest number of workers. It is the official policy of the government to develop the economy and fight poverty through the development of small-scale businesses. Four times SMEDAN promised to rehabilitate the industrial centers and failed. In January 2016, a former director-general of SMEDAN, Bature Masari, said the agency was working toward the upgrade and conversion of its IDCs to enterprise centers and MSME cluster parks in an effort to facilitate the speedy development of MSMEs to enhance economic empowerment and employment generation. He gave the names of the IDCs slated for conversion to include those in Ogun, Ondo, Bauchi, Edo, Kano, Borno, Niger, Cross Rivers, Osun, Rivers, Sokoto, Adamawa State, and Kaduna states. Bature also disclosed that IDCs in Enugu, Abuja, Lagos, Kwara, Plateau, Katsina, Imo, Akwa Ibom, and Taraba States would be converted into Enterprise Zones because of their size to offer common facilities and workspaces to MSMEs in those states. He said SMEDAN was into partnership with Osun and Kano state governments for the redevelopment and upgrade of IDCs in Oshogbo and Tiga respectively where huge funds were being committed by the two-state governments on the upgrade and conversion of the facilities. Unfortunately, when a follow-up visit was made to Kano IDC, there was no sign of any redevelopment. Like other centers, it was deserted and dilapidated. Bature pointed out that while the Tiga IDC was in the process of being converted to a world-class leather cluster park, more than N200 million has so far (by 2019) been expended on the improvement of facilities and rehabilitation of the Oshogbo IDC by the Osun State government under an MOU with SMEDAN. Since then, nothing has happened to the centers, but the empty promises by the government did not stop. This was despite Nigeria's growing unemployment rate, which stood at 23.1% of the workforce in the third quarter of 2018, up from 18.1% a year earlier, according to the National Bureau of Statistics, NBS.

In his message at the opening ceremony of the Lagos Leather Fair, the DG of SMEDAN, Dikko Radda, like his predecessor, said the agency was collaborating with Kano State Government to redevelop an N12 billion IDC in Tiga town of the state into a world-class leather cluster park and training center, but with public-private partnership (PPP) with stakeholders of the leather industry, including potential investors to embrace the project. He spoke through the agency's Director of Engineering, Technology, Innovation and Infrastructure, David Abu Ozigi. That was in June 2017.

Ironically, Kano IDC in Tiga town is currently a grazing field to herders' cattle. There are no signs of any new investment at the center when it was visited in 2019. A year after, June 2018, Radda, again, announced that the Federal Government had put in motion the process of rehabilitating all the IDCs across the country. Radda whose office supervises all the 23 IDCs in Nigeria said the government had plans to transform the centers into world-class enterprise clusters for rapid economic development geared toward job and wealth creation (Akinwale, 2019). He disclosed that SMEDAN in collaboration with the African Development Bank (AfDB) carried out a study of the viability of the centers. The bank, according to him, sponsored the study at a cost of more than $600,000 for the six-month period; the report of which he disclosed had been submitted to the government.

While nothing has been done since then, Radda would also in November of the same year 2018 re-echo that the Federal Government had put in motion the process of rehabilitating all the IDCs. He was leading members of the Senate Committee on Industries on an oversight function to the Owerri Centre, and said the agency had commissioned a study in collaboration with the AfDB on how to rehabilitate the centers. "I believe that by the end of June, the project with the bank will come to a conclusion. What we are waiting for is the submission of the full business scale for six out of the 23 centres," Radda was quoted as saying. Once again, it is reiterated that Kano IDC, which collapsed in 1999, receives only repeated empty promises. It is not an open cattle-grazing land. Mr. Yaro is one of the remaining six workers at the IDC as it is called by the people of Tiga. The center was established in 1982 but suffered a great setback when the Federal Government under former president Olusegun Obasanjo ordered a downsizing of the civil service.

Inside its wide expanse of land were cattle grazing on dried grasses while structures with blown-off roofs stood derelict. Offices were covered in dust; ceiling, doors, and windows were broken. There were no signs of life on the premises. "At the initial stage of my appointment, here was booming but all of a sudden, everything crumbled during 1999 rightsizing and downsizing activities of the Federal Government," Mr. Yaro recalled with nostalgia. "No activity has taken place here since 1999," he added. Asked about the machines at the center, Yaro said, there were machines but they were not functioning because they were all obsolete. According to him, six employees are currently working at the center, including its coordinator whose office is in Zaria, Kaduna State. Due to the peculiarities of Kano for leather production, the IDC was intended to be a hub for leatherwork, but Mr. Yaro said the center which also ought to carry out business appraisal has not witnessed any productive activities for years. This is despite claims by SMEDAN that the Tiga IDC was in the process of being converted to a world-class leather cluster park. That's not the only thing that is wrong with the center, its location is also part of the reasons for its abandonment. The center is located in the outskirts of Kano, making it difficult to access. The location is hundreds of kilometers away from the heart of Kano city.

After spending more than 2 hours searching needlessly for the center's location, the reporter (Akinwale, 2019) got help from an official of the Kano State Chambers of Commerce and Industry who gave a hint of its location. He simply said the IDC is located in Tiga town, but without giving a specific address or a landmark. Before then, officials of Industrial Training Fund (ITF) in the city could not even make sense of the name Industrial Development Centre when asked. None among those asked had heard of it before, likewise artisans along Hadeija road in the city. Situated on Kilometer 1 along Tiga Hydro Electric Power Project, Kano IDC stood forsaken by the roadside. It is a shadow of its old self. The signpost that welcomes visitors has long faded. Only eagle-eyed visitors could see it from afar.

According to Akinwale (2019), farmers are the ones tending the abandoned Rivers State IDC located in Port Harcourt. While the Federal Government continues to brag about rehabilitation, without taking any serious action, those interested in crop production are not allowing its fertile soil to waste away. The center has been long abandoned, overtaken by weeds and parts of it already converted to farms by some people who tend the facility. Apparently, the center has not worked or been used for any industrial purpose in a long while. It was reported that equipment, mostly fabricating machines, installed at the center, were never used until most of them were vandalized, stolen, misappropriated, or became obsolete. This is not surprising because this author, on his nationwide IDC tours in 1994, witnessed similar instances of missing pieces of equipment that were recorded on pre-travel inventory of the center's assets. Questioning and recommending did not yield any corrective actions. Obviously, nothing changed positively between 1994 and 2019. So, what are the prospects for the future. If a systems approach is not instituted and followed, the same pathetic outcomes will happen again.

In Port Harcourt, the roof of the IDC workshops is partly blown off by storm while an old Volkswagen car sits in the middle of the hall. The large workshop is littered with remnants of obsolete wooden and metal works equipment. The entire premises is covered with overgrown grasses. Tired of idleness, staff at the center have stopped coming to work. When there is training to be organized, trainees are usually taken elsewhere due to a lack of appropriate machines for practical work. Unfortunately, this author encountered the same pattern of neglect and ghost workers at many of the IDCs in 1994. The pitiful practice has, obviously, continued since that time.

Akinwale (2019) asked about the procedure to secure a space allocation by an entrepreneur. One of the men taking shelter in the large workshop responded that no one has been given any space in a long time. "We have not received any information from Abuja about the allocation of space to anybody," he said. He insisted that order must come from Abuja before anything can be done. "Here, we cannot do anything, we don't have that power to do anything and nobody has been given any allocation here as you can see," he says. Unfortunately, Abuja, where the official order was to come from, is, itself, in disarray regarding the fate and future of the IDCs. There is an utter lack of coordination across the spectrum of government oversight for the IDCs.

The center has been abandoned for so long that no local seems to know its location in Port Harcourt, including those who live around or operate businesses along NTA Road, Ozuoba, where it is located. Nobody could say the type of activities that take place within the large premises. "The place is called 'Industrial Gate,'" says a middle-aged woman who roasts ripe plantain adjacent to the center, though there are no signs of any industrial activities taking place there at the time. Only a few locals know it as "Industrial Gate"—the large almost faded signpost rests somewhere on the fence near the pedestrian gate, covered by trees nearby which make it hardly noticeable for passers-by. Akinwale (2019) reported that Sam Egwu, a former chairman of the Senate Committee, once likened the center to a mechanic workshop when he led members of the committee on an oversight function visit. He said the visit showed that the IDC only exists in name.

So, where should we go next, following this extensive case study report? If the IDCs are going to be revitalized and refurbished, as the government has claimed several times, a systems approach must be embraced. This will facilitate paying attention to all the facets and nuances involved in running a productive industrial development center. Absence of the use of a sweeping systems framework, the same shortsighted occurrences of the past will repeat again and again. In this regard, it is recommended that the DEJI Systems Model or any of the other systems engineering models be adopted for the future of Nigeria's IDCs.

DEJI SYSTEMS MODEL ANALYSIS OF NIGERIA'S IDCS

The regretful accounts reported above touch this author deeply because of the personal and professional investment of time, effort, and funds in serving as a UNDP consultant to the IDCs of Nigeria in the 1990s. What went wrong? How can the wrong turns be averted in future efforts? Can the IDCs really be revitalized or realigned to support Nigeria's industrialization goals in the prevailing era of digital transformations? This case study shows failure along each and every stage of what DEJI Systems Model preaches.

Design: Although the concept of IDC was sound, it was not thought-through properly to ensure a realistic and sustainable operation. This was constantly mentioned during this author's consultation with UNDP and Nigeria's Ministry of Industry in 1994. In fact, written recommendations were drafted and presented on "Operating Guidelines for IDCs," (Badiru, 1994). Obviously, none of the guidelines were followed in the subsequent years following the author's service. Thus, the IDCs constitute a failure, in terms of the design stage of DEJI Systems Model.

Evaluation: Reading through the testimonies in the case study, it is seen that all the evaluations done all along the way by various government parastatals were done haphazardly or done just on paper without a seriousness to actually implement. Things are often written only to appease government programs and appeal for more government funding. If accountability existed with respect to aligning the design concept with the subsequent stages of a system's approach, more rigorous efforts might have presented the eventual collapse of the IDC concept. Consequently, the IDCs failed, in terms of the evaluation stage of DEJI Systems Model.

Justification: Regardless of the outcomes of the design and evaluation stages, there should always be an explicit focus on justifying the existence and continuation of each IDC at each point of the government's investment. UNDP and other foreign external aids were orchestrated for Nigeria's IDCs in the 1990s. In hindsight, those aids might not have been properly justified due to the noticeable failures in the design and evaluation stages. In fact, in his consultancy report to UNDP in 1994, this author recommended against any further UNDP investment in the IDCs until it could be shown that a serious operating template was in place and being followed. Unfortunately, the response to that particular recommendation was that Nigeria, as a UN member nation, was eligible for the UNDP funding, regardless of the findings and observations at that time.

Integration: The arrant failure of the IDCs in the integration stage of the DEJI Systems Model could not have led to anything else other than what was revealed in the case study. There was no effort made to integrate and align many of the IDC operations to the local environment in which they were located. A particular case is the location of Kano IDC far out in the outskirts of its host city of Kano, to the extent that it is not known by the local residents and could hardly be found even by a dedicated search by an investigative reporter. This misalignment of location makes it difficult for employees, patrons, and clients of the IDCs to engage with the center productively. Thus, failure was the only pathway certain for the IDC, as a result of the failure of integration.

REFERENCES

Akinwale, Yekeen (2019), "INVESTIGATION: Nigeria's Industrial Development Centres in its 'death throes' (Part 1)," https://www.icirnigeria.org/investigation-down-and-out-nigerias-industrial-development-centres-await-bulldozers/, accessed January 27, 2022.

ASCE (2008), The American Society for Civil Engineers, 2013 Report Card for America's Infrastructure, (www.infrastructurereportcard.org), Reston, Virginia, 2013.

Atta, Jacob K. (1981), *A Macroeconomic Mode of a Developing Economy: Ghana*, University Press of America, Washington, DC, 1981.

Badiru, A. B. (1991), "OKIE-ROOKIE: An Expert System for Industry Relocation Assessment in Oklahoma," *Proceedings of the Fifth Oklahoma Symposium on Artificial Intelligence*, Norman, Oklahoma, November 1991.

Badiru, Adedeji B. (1993), *Managing Industrial Development Projects: A Project Management Approach*, Van Nostrand/John Wiley, New York.

Badiru, Adedeji B. (1994), "Operating Guidelines for Nigeria's Industrial Development Centers," *Consultant's Report submitted to Nigeria Ministry of Industry*, Abuja, Nigeria, August 1994.

Badiru, Adedeji B. (2016), *Global Manufacturing Technology Transfer: Africa-USA Strategies, Adaptations, and Management*, Taylor & Francis Group/CRC Press, Boca Raton, FL.

Badiru, Adedeji B., Oye Ibidapo-Obe, and Babs J. Ayeni (2019), *Manufacturing and Enterprise: An Integrated Systems Approach*, Taylor & Francis Group/CRC Press, Boca Raton, FL.

Gharajedaghi, Jamshid (1986), *A Prologue to National Development Planning*, Greenwood Press, New York, NY, 1986.

Hassan, Moshood Ayinde, and Sunday Olawale Olaniran (2011), "Developing Small Business Entrepreneurs through Assistance Institutions: The Role of Industrial Development Centre, Osogbo, Nigeria," *International Journal of Business and Management*, Vol. 6, No. 2, pp. 213–226. DOI:10.5539/ijbm.v6n2p213. Note: Akinwale's investigation was supported by the International Centre for Investigative Reporting (ICIR) and the Ford Foundation.

NAE (2008), *National Academy of Engineering, Grand Challenges for Engineering*, US National Academies, Washington, DC, 2008.

Sahel Publishing (1992), *Industrialization in Nigeria: A Handbook, Nigeria Federal Ministry of Industry and Technology*, Sahel Publishing & Printing Co. Ltd, Lagos, Nigeria.

11 Case Study of DEJI Systems Model® Application to Intelligence Operations

INTRODUCTION

Intelligence gathering and analysis must be integrative with respect to how people think, act, behave, and react. This, essentially, calls for the application of a systems thinking approach. Badiru and Maloney (2016) present a case study that is adapted for this chapter. The field of intelligence is amenable to innovative process modeling. This is particularly useful in assessing terrorist patterns. There are terrorist groups around the world whose primary mission is to inflict as much harm on American interests and citizens as possible. These groups only have to be successful once, as 9/11 taught us, but stopping them requires a 100% success rate by homeland security and national defense efforts. This means having to be right a large percentage of the time in assessing threats and in effectively deploying adequate resources to minimize the risks. This requires robust intelligence gathering and application to execute preemptive strikes. The premise of this paper is that the more systems modeling is applied to an intelligence strategy, the more robust the strategy would be. There will never be sufficient resources to meet every threat. The problem is to deploy the available resources judiciously in the most effective manner so that the overall risk of a terrorist attack is minimized. In this paper, we propose the application of the DEJI (design, evaluation, justification, and integration) systems engineering model to intelligence applications. Specifically, we propose applying the DEJI model in a conceptual framework by using HUMINT (Human Intelligence) to augment SIGINT (signal/sensor intelligence) from a systems perspective.

HOMELAND SECURITY ASPECT

Since September 11, 2001 (9/11), homeland security concerns have dominated the national agenda. Amid unprecedented tragedy, many government agencies rapidly initiated a vast array of security improvements as stop-gap measures to protect critical facilities, transportation systems, and other infrastructure. As the country and the intelligence community (IC) adjust to the "new normalcy," it is logical to question whether

- the right things are being done to deter, detect, prevent, or mitigate future terrorist attacks;
- too much or too little effort, and too many or too few resources, are being applied;
- new technologies will help deter attacks or simply cause the attacker to use a different, but equally effective, means to achieve the same result; and
- all the factors that impinge on national security are being factored into intelligence decisions.

These are logical questions. No one can state definitively what should or should not be done to detect and prevent terrorist attacks, but some level of consensus decision-making will have to occur. It has been shown that effective homeland security is not just a matter of technological gadgetry but also of an effective high-fidelity decision process. It is the position of this paper that a systems-based adaptive risk-based approach can help improve decisions, particularly decisions that must be based on uncertain information. Systems risk-based decision-making (SRBDM) can help decision-makers (users) address these questions and a wide range of other related questions.

For example, the U.S. Navy (Navy) is using a risk-based approach to evaluate and implement an interdependent suite of antiterrorism (AT) capabilities aimed at increasing the Navy's ability to deter, detect, and respond to terrorist threats. While many capabilities already exist and others are being developed, the Navy must make decisions on how to allocate resources in a resource-constrained environment to best manage the risks associated with security threats. The Commander, Navy Installations (CNI) has sponsored efforts to link the results of a risk-based model with classic operations research modeling to help optimize the allocation of limited resources among many anti-terrorist capabilities. The prototype Navy model (at that time) was not comprehensive and did not adequately cover all decision factors, but it did demonstrate the feasibility of developing an adaptive risk-based resource-allocation decision tool for homeland security application. Although the AT capabilities were considered individually, significant dependencies exist between some capabilities due to the integrated performance expected from the systems that will be implemented to reduce the overall threat of a terrorist attack. For example, certain robust command and control actions rely heavily on information management and display infrastructure, and on communication capabilities between various Naval and civilian agencies. Therefore, the benefits associated with improved command and control systems cannot be fully realized without also addressing infrastructure needs related to information management and communication. This presents the following challenges:

1. How to directly incorporate interdependencies into a risk model?
2. How to determine the benefit (reduction in risk) that might be derived if a particular capability is only partially implemented?

3. How to change the assessment of overall risks and adjust strategies given the two-player nature of intentional attacks?
4. How to incorporate intelligence (both HUMINT and SIGINT) into AT decision-making?
5. How to ensure an integrative continuity of the AT strategy?

TECHNICAL BACKGROUND

Researchers all over the nation are developing, testing, and integrating advanced signal-processing, image-processing, and data-processing technologies for high-fidelity sensing systems. Improved technologies will increase reliability and accelerate the speed at which data is transmitted from sensing systems to humans, who monitor and analyze the data. Badiru and Maloney (2016) present a conceptual framework for an innovative application of a systems engineering model to intelligence operations. Intelligence on terrorist activities will always be incomplete and imperfect. Intentional planned attacks involve two opponents with competing objectives. Each opponent will consider the options and objectives of his adversary in formulating a strategy and will change his strategy based on what his opponent does or what he expects him to do. Although it is clear that some terrorists are willing to die while attacking a target, there is no evidence to suggest that a terrorist will deliberately attack a target if he perceives that he will fail in the attempt. Therefore, making a target invulnerable to a particular mode of attack will not necessarily put the terrorist out of business; rather, it may cause him to consider other targets or modes of attack. Therefore, taking action to reduce a specific threat may increase the likelihood that an alternative target will be attacked. These are all issues that must be addressed from a systems perspective within the realm of effective human decision-making, not just from a technical assets point of view, as embodied in the DEJI model (Badiru, 2012, 2014).

Decision-makers must be aware that terrorists will adjust attack modes and adopt strategies that will exploit vulnerabilities in a dynamic manner. Actions designed to reduce vulnerability to a specific attack mode must not only be considered to ensure that they are effective; their impact on other scenarios must also be considered. Myopic focus on safeguards for a specific target or attack mode could simply shift the risk to other targets or attack modes and could actually increase overall risk. The research question is posed as follows:

Can an effective systems-based methodology be devised that will effectively **Design, Evaluate, Justify,** and **Integrate** the trade-offs between risk and costs, acknowledge the interdependencies in resource-allocation decisions and the integrated nature of systems execution, respond in a timely manner to counter strategies perceived to have been made by the terrorist adversaries, take into account the imperfect and incomplete nature of intelligence on terrorist activities, and effectively guide resource allocation decisions so as to minimize the overall threat of an intentional attack?

In the approach of this paper, we recommend using the systems-based DEJI model, which has been applied to a variety of practical problems (Badiru, 2012, 2014)

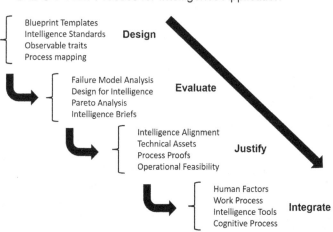

D-E-J-I Flow Process for Intelligence Application

FIGURE 11.1 DEJI Systems Model for intelligence analysis.

dealing with designing, evaluating, justifying, and integrating problem parameters and factors. Figure 11.1 illustrates the basic structure of the DEJI model.

The research problem has two components:

- the *risk-assessment* part in which benefits (defined as reduction in the risk of attack) are moving targets that must be continuously re-evaluated based on current intelligence and dynamically fed into a capital budgeting model for strategy development and adjustment.
- the *resource allocation* part in which decisions are generated as to which capabilities should be implemented, and at what level, in order to minimize the overall threat based on the most recent systems risk assessment.

SYSTEMS METHODOLOGY

Given the wide range of attack scenarios (i.e., possible terrorist targets and attack modes), and the uncertainty associated with each scenario, the benefit provided by a particular resource allocation can be difficult to predict. However, we can establish a structured methodology to evaluate the risk benefit for each capability based on how each capability might change the threat, target vulnerability, or consequences (TVC) for each scenario.

DESIGN SECTION

Under the DEJI Systems Model approach, the design aspect relates to the resource allocation model as formulated in the following large-scale nonlinear mixed binary integer programming problem:

Let:

I = set of counterterrorism capabilities

T = set of funding cycles (e.g., fiscal years)

J = set of funding sources $\{1, 2, 3, 4, \dots\}$

e.g., funding source: 1 = capital works; 2 = research & development; 3 = procurements; 4 = operations/maintenance

x_{ijt} = proportion of full funding allocated to capability i from funding source j in time period t

$\phi(x_{i4t})$ = utility function; expressed as the % benefit derived as a function of funding level (% of total funding) in time t – applied only to operations/maintenance funding source

$$y_{it} = \begin{cases} 1, & \text{if capability } i \text{ is deployed at time } t \\ 0, & \text{otherwise} \end{cases}$$

$$i \in I \quad j \in J \quad t \in T$$

c_{it} = reduction in risk (threat) achieved by fully implementing capability i in time t.

Objective : *Maximize* the reduction in the risk of a terrorist attack.

$$\max \xi = \sum_i \sum_t c_{it} y_{it} \phi(x_{i4t})$$

Conditions:

1. Appropriation limits (budgets from each funding source cannot be exceeded)
2. Dependencies across funding sources (e.g., sequencing of spending)
3. Dependencies across time periods (e.g., sustaining funding commitments)
4. Dependencies across capabilities (e.g., funding prevention before detection)
5. Contingencies (either/or, if a, then b)
6. Deployment (only allocate funds to deployed capabilities)
7. Bounding (force minimum funding levels, if appropriate)
8. Structural (e.g., linearization, binary)

The proposed model consists of the modules shown in Figure 11.2.

EVALUATION SECTION

The capital budgeting class of problems, including uncertainty and risk, has been investigated from numerous perspectives. Recent work includes the use of fuzzy numbers to estimate uncertain returns, a modified weighted average cost of capital

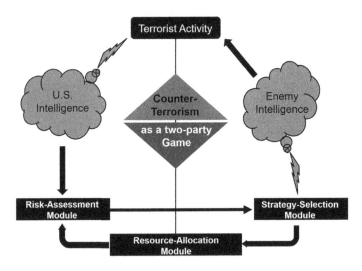

FIGURE 11.2 DEJI Systems Modeling for intelligence problem abstraction.

methodology to project returns, pooling of risks across multiple projects, analytic hierarch process (AHP) as a decision framework, zero-one integer programming to accommodate sequencing decisions in a dynamic environment, an integrated approach that combines risk management with capital budgeting, a methodology for selecting projects in high-risk R&D environments, goal programming for decision-making in an uncertain environment, a generalized dynamic capital allocation methodology using distortion risk measures, sensitivity analysis as a tool applied to capital budgeting under uncertainty and risk, and game theory combined with Monte Carlo simulation for timing resource allocations in a homogeneous commodity market.

The desired methodology will allocate resources sequentially across competing highly interdependent projects, where each project may be partially funded, with nonlinear utility functions that are dependent on the counter-moves of an intelligent adversary. A model to allocate resources to combat terrorism must address all these issues from an overall systems perspective. Developing such a model requires the integration of mathematical programming, stochastic processes, risk assessment, and classical game theory.

1. There is an urgent need for decision tools that will reduce the threat, help decision-makers to spend public money more wisely, and rapidly respond to changes in the threat due to actions by the terrorists.
2. The integrative nature of the proposed model represents an application of established research to new areas of expertise.
3. The use of a decision model that directly links resource allocation strategy dynamically with counter-moves by a terrorist adversary, factoring in the imperfect nature of intelligence information and the political process, has largely been untested.

Because acts of terrorism can be vastly different, widespread, and involve ever-changing methods, analysis of information needs to be performed every step of the way. Those who gather intelligence need to realize that some information is more vital to national security and the analysts themselves need to take into account the very human aspect of terrorism. Humans are irrational at best, especially when angry and trying to get a point across. Analysts in today's world need to be trained in the art of analysis, not just specific topics or regions. Analysts nowadays must have a global perspective that was not needed during the predictable days of the Cold War. Additional funding to analysis is vital. This funding would allow additional analysts to keep an eye on small threads linking bits of information to potential terrorist attacks. While the large amount of information possessed by the IC is impressive, national security interests are only protected if that information is transformed into intelligence.

We are no longer in a Cold War world. In our world, terrorist organizations are proud of their blatant disregard for humanity. The IC can no longer fund specialists like they could in the mid to late 20th century. When the United States had one large, known enemy, it was possible for analysts to pick a specialty and stick with it their entire career. Now it is nearly impossible to have specialists on every threat to the United States. From Russia, China, and North Korea, we face states that would like to see the United States taken down a peg; from the Islamic State of Iraq and the Levant, Hezbollah, and Boko Haram we face non-state actors with no limitations on their cruelty and barbarism.

JUSTIFICATION SECTION

Assuming a set of predetermined strategic scenarios, risk-based payoff values will be developed and evaluated using a modified risk-assessment procedure and appropriate statistical procedures. The best strategy can be found by solving a mixed (randomized) strategy game problem based on the payoff matrix obtained from the risk-assessment module. The optimal solution of a linear programming (LP) formulation will represent the probabilities for each strategy to return the best-expected payoff value. Based on the optimal solution from the strategy selection module, the predetermined strategies can be prioritized. The objective of the proposed resource allocation model is to maximize the reduction in the risk of a terrorist attack and can be formulated as the large-scale nonlinear mixed binary integer programming problem.

The money it takes to build a new satellite is astronomical compared to the money it takes to hire additional analysts for different intelligence agencies. As stated by Best (2015), "Unfortunately, sophisticated political and social analysis is often not emphasized in intelligence agencies, especially within the Defense Department, that are focused on technical collection and direct support to operational commanders." The U.S. government is setting the IC up to fail should policy makers not fund what the IC deems necessary. The cost is very low when it comes to linking intelligence agencies and allowing them to share their information and analysis. The way the IC had grown to be so bureaucratic and have immense tangles of red tape can be reversed only if policy makers decide that information sharing between agencies is as vital as many IC members claim.

INTEGRATION SECTION

This section is presented as a hypothetical case example of the importance of integration in intelligence analyses.

CASE EXAMPLE: MORE ATTENTION TO HUMAN INTELLIGENCE IN ORDER TO AUGMENT SIGNALS INTELLIGENCE

The intelligence community, including the armed forces, has focused the majority of their research, time, and money on signals intelligence (SIGINT) while leaving Human Intelligence (HUMINT) without sufficient attention and funding. Though SIGINT worked wonderfully during the Cold War to decode and decrypt Soviet messages, the developing world has seen a rise in non-state actors threatening the United States. These non-state actors utilize the exponential growth of social media and the Internet. Because there are entirely too many pieces of information circulating every day, vital information can fall through the cracks. SIGINT cannot, and should not, be responsible for keeping track of all signals. Therefore, HUMINT should come back into the spotlight. The intelligence community should put a renewed focus on quality over quantity.

Many countries around the world have developed high-performing and gainful human intelligence agencies. Though these nations tend to use tactics considered inhumane or illegal, places like Russia and Israel have found a way to gather relatively secure sources in places where SIGINT is no longer the best option. The United States needs human intelligence in order to be better prepared to deal with small terrorist cells and lone-wolf attacks. While the United States values human rights and should not go to cruel measures to secure information, the United States needs to understand the importance of intelligence "boots on the ground."

In many ways, SIGINT intrigues people. Congress is fascinated by new technology and likes to physically see what they are funding. This emphasis on SIGINT, however, is not always the most cost effective. Human intelligence, though occasionally subject to manipulation and deception, can provide data and information that signals intelligence cannot. A human being can see if a person looks worried while talking to different individuals or takes extra caution crossing a certain stretch of the road. A human being can detect when a voice seems weary or cautious. A human being can begin to bond and build relationships with people who hold vital information. While reading communications and listening to recordings is wonderful for quantitative information, the United States is in dire need of qualitative information if there is any hope in staying two steps ahead of her adversaries. As the Head of Intelligence Collation Management at the North Atlantic Treaty Organization's mission in Sarajevo, Palfy (2015) said it best when he stated, "... increased collection does not necessarily or automatically lead to better intelligence outcomes."

We can look to history to compare and contrast SIGINT and HUMINT. During the Cold War, missions like the Bay of Pigs had many pictures and technical information about Soviet intervention in Cuba, but there was very little focus on HUMINT. Had the United States noticed that the Cuban people did not want to overthrow Castro, they may have been spared the embarrassing blemish on the Kennedy administration.

The implementation of this shift in concentration would not likely be difficult. The most time-consuming and difficult step in the process would likely be funding approval from Congress. The Congress people would likely be unhappy to cut funding to SIGINT because, in many cases, creating technology used for SIGINT brings in money and jobs to their constituents back home. As long as Congress approves the reassignment of funds, integrating this new policy should be smooth. SIGINT should continue to be funded, of course, because of its vital role in the information age.

CONCLUSION

This paper has presented a conceptual framework for the application of a systems approach to addressing systems challenges. Although no specific problem is tackled in the paper, the framework can give readers in the intelligence community an expanded idea of how to design, evaluate, justify, and integrate intelligence strategies. In recent years, sensor-based systems have increased in development and applications. The variety and diversity of HUMINT and SIGINT systems necessitate the application of a systems approach. This paper has introduced the application of the DEJI model by proposing the use of HUMINT as an integrated augmentation of SIGINT.

REFERENCES

Badiru, A. B. (2012), "Application of the DEJI Model for Aerospace Product Integration," *Journal of Aviation and Aerospace Perspectives (JAAP)*, Vol. 2, No. 2, pp. 20–34, Fall 2012.

Badiru, A. B. (2014), "Quality Insights: The DEJI Model for Quality Design, Evaluation, Justification, and Integration," *International Journal of Quality Engineering and Technology*, Vol. 4, No. 4, pp. 369–378.

Badiru, Adedeji B., and Anna E. Maloney (2016), "A Conceptual Framework for the Application of Systems Approach to Intelligence Operations: Using HUMINT to augment SIGINT," *American Intelligence Journal*, Vol. 33, No. 2, pp. 41–46.

Best, Richard A. (2015) "Intelligence and U.S. National Security Policy," *International Journal of Intelligence and Counter Intelligence*, Vol 28, No. 3, pp. 449–467, DOI: 10.1080/08850607.2015.1022460

Palfy, Arpad (2015) "Bridging the Gap between Collection and Analysis: Intelligence Information Processing and Data Governance," *International Journal of Intelligence and Counter Intelligence*, Vol. 28, No. 2, pp. 365–376, DOI: 10.1080/08850607.2015.992761

Index

Printed in the United States
by Baker & Taylor Publisher Services